T0205385

Springer Theses

Recognizing Outstanding Ph.D. Research

Aims and Scope

The series "Springer Theses" brings together a selection of the very best Ph.D. theses from around the world and across the physical sciences. Nominated and endorsed by two recognized specialists, each published volume has been selected for its scientific excellence and the high impact of its contents for the pertinent field of research. For greater accessibility to non-specialists, the published versions include an extended introduction, as well as a foreword by the student's supervisor explaining the special relevance of the work for the field. As a whole, the series will provide a valuable resource both for newcomers to the research fields described, and for other scientists seeking detailed background information on special questions. Finally, it provides an accredited documentation of the valuable contributions made by today's younger generation of scientists.

Theses are accepted into the series by invited nomination only and must fulfill all of the following criteria

- They must be written in good English.
- The topic should fall within the confines of Chemistry, Physics, Earth Sciences, Engineering and related interdisciplinary fields such as Materials, Nanoscience, Chemical Engineering, Complex Systems and Biophysics.
- The work reported in the thesis must represent a significant scientific advance.
- If the thesis includes previously published material, permission to reproduce this must be gained from the respective copyright holder.
- They must have been examined and passed during the 12 months prior to nomination.
- Each thesis should include a foreword by the supervisor outlining the significance of its content.
- The theses should have a clearly defined structure including an introduction accessible to scientists not expert in that particular field.

More information about this series at http://www.springer.com/series/8790

Azadeh Mirabedini

Developing Novel Spinning Methods to Fabricate Continuous Multifunctional Fibres for Bioapplications

Doctoral Thesis accepted by
the University of Wollongong, Australia

 Springer

Author
Dr. Azadeh Mirabedini
Advanced Technology Centre/Faculty of
  Science, Engineering and Technology
Swinburne University of Technology
Melbourne, VIC, Australia

Supervisors
Dr. Gordon G. Wallace
University of Wollongong
Wollongong, NSW, Australia

Dr. Javad Foroughi
ARC Centre of Excellence for
  Electromaterials Science (ACES)
  Intelligent Polymer Research Institute
  (IPRI)
University of Wollongong
Wollongong, NSW, Australia

ISSN 2190-5053   ISSN 2190-5061 (electronic)
Springer Theses
ISBN 978-3-030-07023-6   ISBN 978-3-319-95378-6 (eBook)
https://doi.org/10.1007/978-3-319-95378-6

Printed on acid-free paper

This Springer imprint is published by the registered company Springer Nature Switzerland AG
The registered company address is: Gewerbestrasse 11, 6330 Cham, Switzerland

This work is dedicated to my dearest husband and daughter, Mr. Saber Mostafavian and Ms. Baran Mostafavian and my parents, Mrs. Heshmat Mirabedini and Mr. Hassan Mirabedini for their endless encouragement, love and support.

Supervisor's Foreword

Unlike a cut that heals, the central nervous system has limited capacity to fully recover damaged or severed connections in the event of an injury. These nerves, thus, need to be regenerated or repaired, synthetically. Current state-of-the-art microelectrodes generate an electrical field in the tissue using metallic implants to elicit or suppress neuronal action potentials, though they often suffer from biocompatibility issues. In contrast, core–sheath fibres seem to provide a more appropriate, biocompatible replacement for damaged nerve fibres. "Developing Novel Spinning Methods to Fabricate Continuous Multifunctional Fibres for Bioapplications" is probably the first book dealing with optimisation of coaxial solution spinning processes for the development of multifunctional biocompatible fibres. This book represents the development of innovative methodologies in the fabrication of multilayered fibres mainly for use in biological applications. The main concept of this work is to use naturally occurring hydrogel platforms as non-conducting, hydrophilic and biocompatible materials as the sheath component to promote cell adhesion whilst incorporating the conducting elements in the core of the fibres.

This book is of particular significance because it presents a careful study of the concurrent controlling parameters affecting the wet spinning of multicomponent fibres. Prior to this study, only a few researches have described the successful fabrication of either coaxial or triaxial fibres *via* the novel application of the wet-spinning methodology. The author has successfully adapted some hybrid material formulations and details what is required to wet-spin these into layered fibres. Each experimental chapter has two parts: one detailing the optimisation of the wet-spinning process for the specific spinning solutions under investigation, while the other assessing and discussing different properties of the fibres produced. Consequently, as the reader of this book, you will learn about the important factors involved in the wet-spinning practice including optimising the solvent selection, understanding the rheological properties of the spinning solutions, selection of the appropriate coagulation bath and choosing the correct injection rate(s), as well as many other processing considerations. Additionally, one can become familiar with evaluation methods for the core–sheath fibres as they relate to bioapplications.

The book targets and has relevance across a variety of disciplines including materials scientists, biologists, electrochemists, neurologists, surgeons and students. The author should be duly acknowledged for her diligence and effort in bringing this material to the attention of this broader audience and we are confident that our readership finds this book timely and informative.

Wollongong, Australia   Dr. Gordon G. Wallace
April 2018   Dr. Javad Foroughi

Abstract

Developmental work in the field of multifunctional hybrid fibres has revealed a number of characteristics that promise great benefits to their possible use in a broad range of devices and applications including tissue scaffolds and implantable electrodes as well as the accessory energy storage devices necessary to provide power to implantable devices and smart garments. There has, therefore, been much interest and many attempts to produce lightweight, foldable and electroactive multiaxial fibres. The main aim of this thesis is to establish a wet-spinning process to develop three-dimensional coaxial and triaxial electroactive fibres. Using a coaxial wet-spinning method that takes advantage of the inherent electroactivity in a conductive core, we also aim to improve the fibres' mechanical, biocompatibility and cell adhesion properties by using an appropriate biomaterial for the surrounding sheath, opening up the possibility for its use in many biomedical applications. Although some success has been achieved *via* the use of either electrospinning or coating approaches, only a few studies have reported the successful fabrication of either coaxial or triaxial fibres *via* novel application of wet-spinning methodology to the best knowledge of the author. This poor rate of success may be due to the complexity of this method in terms of the number of concurrent solution and processing parameters that need to be optimised and controlled in order to achieve the successful (and consistent) formation of a core–sheath structure in a coagulation bath. Wet spinning has the advantage that it produces individual, collectable fibres that can be drawn out and tested for their mechanical properties. Electrospinning can produce extremely fine fibres in the form of a non-woven mat; however, the mechanical testing of individual fibres is not feasible. There have been also reported on difficulties involved with preparation of fibres using already charged polymer backbones *via* electrospinning wherein a stable jet cannot be achieved and no nanofibres will form, although single droplet may be achieved (electrospray). In addition to this, there is a limitation in choosing the maximum concentration for a given solution by which it could flow. Consequently, the molecular weight and the concentration as long as a solvent with the necessary volatility which can be spun this way are within a certain range. Thus, although this method shows a lot of promise, these restrictions are placed on the spinnability of certain polymers by solution parameters like viscosity and surface charging.

As a result of this research, the production of continuous core–sheath fibres has been achieved using a variety of materials. Hydrogels have been used as the sheath components since they provide mechanical and structural properties that mimic many tissues and their extracellular matrix. Organic conductors such as graphene and intrinsically conducting polymers have been useful in the creation of electrical pathways within the inner core where they are able to provide safe electrical stimulation of the surrounding tissue—enhancing the electro-cellular communication process—whilst also avoiding undesirable chemical reactions and cell damage.

The initial coaxial wet-spinning process has enabled the development of coaxial hydrogel fibres for the first time without the use of a template. The conditions necessary to achieve optimal properties in the chitosan–alginate core–sheath fibres, as well as their physical, biocompatibility and release properties are investigated. Subsequently, the incorporation of graphene conductors into the core material was studied in order to optimise the electroactivity of the coaxial fibres. Use of large graphene oxide sheets has enabled the use of a wet-spinning protocol which can produce strong fibres that are easily converted to electrically conducting graphene fibres *via* treatment with a non-toxic L-Ascorbic acid chemical reducing agent. Further work led to improvements in fibre properties with the substitution of an intrinsically conducting polymer as the core material. Using Chit-PEDOT:PSS fibres as a template, triaxial fibres were fabricated *via* the use of a chemical vapour deposition technique to facilitate the growth of PPy onto the fibre surfaces opening up potential application for their use in biobatteries.

In conclusion, a range of multiaxial fibre types have been developed which show characteristics that lend themselves to potential use in a broad range of applications such as actuators, neural implants, microelectrodes and biomedical devices. The main stated aim of this thesis, to prepare multifunctional hybrid three-dimensional structures, was achieved using a variety of material compositions and properties *via* wet-spinning processing methods. The microfibres produced in this way may find uses in the regeneration of functional tissue scaffolds that mimic muscle fibres, blood vessels or nerve networks.

Preface

With the ever-increasing demand for suitable replacements and organ transplantation, tissue engineering has drawn increasing attention due to its ability to create three-dimensional (3D) structures using biomaterials. State-of-the-art developments in the field of implantable microelectrodes have shown potential for their use in electrical therapies such as auditory implants, deep brain and spinal cord stimulators as well as vision prostheses. Electrodes made out of number of materials especially metals have been conventionally used in this regard; however, they suffer from compatibility issues, such as their surface stiffness, which could trigger chronic biological responses. Using organic conductors and hydrogels, multifunctional hybrid fibres may be produced which can be used as a basis for the manufacture of lightweight, foldable, yet robust electroactive components which are more bio-favourable and thus have potential use not only as implantable electrodes but also in the energy storage systems required to power them, as well as smart garments. It is known that the geometric and structural aspects of anisotropy of fibres result in extraordinary axial properties. Forming and assembling fibres to fabricate 3D constructs would enable us to fully utilise the unique mechanical and physical properties of all involved components at a macroscopic level compared to a bulk material.

Among the many production techniques available, advanced fibre processing, such as coaxial and triaxial spinning of natural and synthetic polymers, has attracted a great deal of attention because the basic core–sheath structure provides a mechanism for the safe electrical stimulation of tissue whilst also avoiding undesirable chemical by-products and subsequent cell damage. Current state-of-the-art microelectrodes generate an electrical field in the tissue using metallic electrodes to elicit or suppress neuronal action potentials. However, the biomechanical and structural mismatch between current electrodes and neural tissues remains a challenge for neural interfaces. This is through the development of advanced fabrication technologies that improvement can be made more widely across the electrical, mechanical and biological properties of electrodes. Encapsulating the conductor cores with a more bio-friendly coating allows for a versatile system for producing devices with appropriate mechanical, chemical and biological properties that can

mimic the native extracellular matrix, better supporting cell growth and mainte-
nance. This thesis presents a novel fabrication method using a wet-spinning process
that allows for the routine production of multifunctional coaxial biofibres that take
advantage of the electroactive properties of a conductive core whilst also promoting
good cell growth and biocompatibility *via* the use of bio-friendly material in the
sheath.

Thesis Outline

Biomaterials are an exciting group of materials being used today by many
researchers in a variety of biomedical applications. As clear from the term, "biomaterials" is a combination of two words of 'bio' and 'materials' which mainly
deals with engineering novel structures with appropriate characteristics to enhance
the interaction with the living cells.

This thesis described the development of three-dimensional electroactive fibres
using a novel coaxial wet-spinning approach from organic conductors in combination with non-conducting hydrogel polymers to be utilised in a diverse range of
applications including implantable electrodes as well as the energy storage systems.
For one, a novel coaxial wet-spinning method was employed for the development
of coaxial hydrogel fibres. Then, insertion of organic conductors as the core
component in coaxial fibres was investigated to create electroactivity. Using a
chemical vapour deposition (CVD) method, the second thin conductive layer of
polypyrrole (PPy) was grown on coaxial fibres leading to the development of
triaxial fibres. The fabricated structures were tested for cytocompatibility as well as
being utilised as a support for cellular adhesion and growth. This work is sum-
marised in the sections as follows:

In Chap. 1, the significance of engineering three-dimensional fibres was highlighted. Second, the involved hydrogels and organic conductors are introduced and
explained briefly with an overall emphasis on their usage for bioapplications
together with some of their selective applications to date. Then, some of the previously established fabrication methods are described, briefly along with a final
focus on the novelty of the fabrication approach applied in this study. The chapter
concludes with some of the potential applications of as-prepared fibres.

Chapter 2 provides an overall insight into the general experimental fabrication
methods as well as the characterisation analyses. However, most of the methods are
explained in detail in the following chapters.

Chapter 3 describes a one-step coaxial wet-spinning approach for development
of biopolymeric continuous core–sheath fibres, with an inner core of chitosan and
alginate as the sheath, for the first time without using a template. The first step was
to spin chitosan and alginate single fibres, separately to enhance and optimise
spinning conditions. The necessary conditions to achieve optimal properties of the
core–sheath fibre were also studied. The physical and mechanical properties of solid
and coaxial fibres were examined. Release profiles from the coaxial fibre were

determined using toluidine blue as a model component. It was indicated that integrating of these polymers with two totally different natures resulted in a creative and innovative fibre fabrication leading to a robust and superior hybrid fibre with higher cell adaptability, improved mechanical, thermal and swelling properties.

Chapter 4 investigates the incorporation of graphene into the conducting component used in the core. In this study, liquid crystal graphene oxide was used due to its high flexibility it provides in case of spinnability as well as the ease of reducing it into a conductive material. One-step continuous wet spinning to produce multi-functional coaxial biofibres of chitosan/graphene oxide in a basic coagulation bath is demonstrated. Spinnability of coaxial fibres as well as their characterisation for several physical and electrochemical (EC) properties is investigated.

In Chap. 5, a novel facile coaxial wet-spinning fibre production method is reported followed by a CVD technique in an attempt to develop a triaxial fibre as a potential battery device. Coaxial fibres of Alginate/poly (3,4-ethylenedioxythio-phene) polystyrene sulfonate (abbreviated as Alg/PEDOT:PSS) and Chitosan/PEDOT:PSS (shortened as Chit/PEDOT:PSS) fibres were spun into a bath of aqueous calcium chloride and sodium hydroxide (NaOH) solution, respectively. The rheological properties of spinning solutions including concentration and viscosity are studied and optimised as most critical parameters in the formation of coaxial fibres. CVD is utilised to create the second conductive layer of polypyrrole on the surface of Chit/PEDOT:PSS fibres. The mechanical, EC and physical behaviours of as-prepared fibres are studied using different analysis methods. It is worth pointing out that cytotoxicity and cell adhesion of the fibres were tested using different types of cells via live/dead cell imaging and calcein staining. The interactions of the cells with fibres were determined for multiple cell lines by looking at the adhesion and proliferation of the cells seeded on the fibres.

Chapter 6 concludes this thesis providing final summary and recommendations for future work. It is also worth mentioning that each chapter contains a short introduction, experimental methods, followed by results and discussion, and a final conclusion section.

Melbourne, Australia   Azadeh Mirabedini

Parts of this thesis have been published in the following Journal articles:

1. Azadeh Mirabedini, Javad Foroughi, Gordon G. Wallace, "Developments in Conducting Polymer Fibres: From Established Spinning Methods toward Advanced Applications", **RSC Advances 2016,** 6(50), 44687–44716.
2. Azadeh Mirabedini, Javad Foroughi, Brianna Thompson, Gordon G. Wallace, "Fabrication of Coaxial Wet-Spun Graphene–Chitosan Biofibers**", **Advanced Engineering Materials 2015**, 18(2), 284–293.
3. Azadeh Mirabedini, Javad Foroughi, Tony Romeo, Gordon G. Wallace, "Development and Characterisation of Novel Hybrid Hydrogel Fibres", **Macromolucular Materials Engineering 2015**, 300(12), 1217–1225.

Conference Proceedings

1. Javad Foroughi, Azadeh Mirabedini, Carbon Nanotube Yarn as Novel Artificial Muscles, 4th Nano Today Conference, 6–10th December 2015, Dubai, United Arab Emirates.
2. Azadeh Mirabedini, Javad Foroughi, Gordon Wallace, Investigation of Electrochemical Actuation Properties of Chitosan-based Novel Microstructures, Asian Textile Conference (ATC-13), 3–6th Nov. 2015, Deakin University, Geelong, Australia.
3. Azadeh Mirabedini, Javad Foroughi, Gordon Wallace, Fabrication Of Multifunctional Coaxial Fibres For Bioapplications, Diamond and Carbon Materials (DIAM), 6–10th September 2015, Bad Hamburg, Germany.
4. Azadeh Mirabedini, Javad Foroughi, Gordon Wallace, Walking on a thin edge, Coaxial/triaxial spinning is a high wire act, 12–15th July 2015, Gold-coast, Australia.
5. Azadeh Mirabedini, Shazed Aziz, Rodrigo Lozano, Javad Foroughi, Gordon G. Wallace, 10th Annual International Electromaterials Science Symposium, Hydrogel-based Twisted Fibre Actuators; looking into Physicochemical/biological properties 12–14th February 2014, University of Wollongong, Australia.
6. Javad Foroughi, Azadeh Mirabedini, Novel biopolymer fibres for biomedical application, New Zealand Controlled Release Society (NZCRS) Workshop, 22–24th November 2014, University of Auckland, New Zealand.
7. Azadeh Mirabedini, Javad Foroughi, Gordon Wallace, ANN Early Career Workshop (ACR), A Novel method for fabrication of wet-spun conductive multifunctional coaxial fibres, 10–11th July 2014, University of Technology Sydney(UTS), Australia.
8. Azadeh Mirabedini, Javad Foroughi, Gordon G. Wallace, "Workshop on Materials for Drug Delivery", 11th February 2014 University of Wollongong, Australia.

9. Azadeh Mirabedini, Javad Foroughi, Brianna Thompson, Sina Jamali, Gordon Wallace, Multifunctional coaxial electroactive fibres, 9th Annual International Electromaterials Science Symposium, 12–14th February 2014, University of Wollongong, Australia.
10. Azadeh Mirabedini, Javad Foroughi, and Gordon G. Wallace, A Novel Wet-spinning Method to Produce Triaxial Fibres, ACES Full Centre Meeting & Ethics Workshop Program, 11–12th June 2013, Hobart, Australia.
11. Azadeh Mirabedini, Javad Foroughi, Gordon Wallace. Novel Approach To Produce Core-Sheath Biopolymers Fibres, 8th Annual International ARC Centre of Excellence for Electromaterials Science (ACES) Symposium, 13–15th February 2013, University of Wollongong, Australia.

Manuscripts in Progress

1. Azadeh Mirabedini, Shazed Aziz, Rodrigo Lozano, Javad Foroughi, Geoffrey Spinks, Gordon Wallace, Bioinspired Moisture-Activated Torsional Muscles from Coiled Chitosan Fibres.
2. Azadeh Mirabedini, Holly Warren, Javad Foroughi, Gordon Wallace, Development of One-dimensional Fibre Bio-batteries.
3. Azadeh Mirabedini, Nicholas Apollo, Javad Foroughi, David J. Garrett, Gordon G. Wallace, Novel Twisted Core-sheath Yarns as Deep Brain Stimulation Electrodes.

Acknowledgements

I would like to thank my supervisors Prof. Gordon Wallace and Dr. Foroughi for supervision throughout this work. I would like to express my sincere gratitude to Gordon for establishing such a remarkable research facility which enabled me to become acquainted with the advanced fabrication and characterisation techniques and gave me the opportunity to grow. And my special regards go to Javad who was more than a supervisor for me, a real mentor and above all, a good friend. He guided, encouraged and supported me throughout my Ph.D. course.

I'm grateful to Prof. Geoffrey Spinks for providing the background I lacked at different stages of the project as well as correction of my paper. I would like to thank Mr. Tony Romeo and Dr. Mitchell Nancarrow at Electron Microscopy Centre for their technical assistance with SEM and optical microscopes. Also thanks to Dr. Anita Quigley, Dr. Brianna Thompson and Mr. Rodrigo Lozano for their collaboration in cell culture studies. Technical advice from Dr. Rouhollah Ali Jalili as well as provision of graphene oxide dispersion is gratefully acknowledged. I highly appreciate Dr. Seyed Hamed Aboutalebi, Mr. Sepehr Talebian and Mr. Ali Jeirani for their occasional help and advices during my Ph.D. I would like to appreciate Dr. Holly Warren for her training on working with impedance and electrical conductivity measurement.

Thanks to all my colleagues and friends in AIIM, who are still here or left for new adventures including Hamed, Stefan, Aynaz, Sina J., Sepehr, Fahimeh, Dharshika, Shazed, Joseph, Rodrigo, Phil, Willo, Dorna, Leo, Mohammad Javad, Danial, Sina N., Alex, Mark, Maryam and Benny who made my Ph.D. journey much more enjoyable.

There is a quote that says, 'Family is not an important thing; it's everything!' There would be no truer words to describe my opinion. I'm very lucky to have a loving and supporting family to give me enough strength to continue the challenging task of doing a Ph.D. Thank you for being who you are. To my wonderful partner, soul mate and best friend, Saber; I could not reach many of my ambitions successfully without your loving heart, caring nature and absolute support. Thank you for making my world an even more beautiful place. To my beautiful and adorable little angel, Baran; you made my heart full of hope and joy with your comforting look, generous smile and sweet words from the very first day you entered my life.

To my parents, Heshmat and Hassan; I cannot thank you enough for everything you have done for me; you have always believed in me, taught me the importance of having integrity, dignified life, work ethic and self-respect and dedicated all the past years to me, my happiness and goals and my brother Nima, my very first best friend, for your continuous love, kindness and backing. Thank you all for everything that I am.

Contents

1 Introduction and Literature Review 1
  1.1 Introduction 1
    1.1.1 Material Considerations for Biomedical Applications 1
    1.1.2 Fabrication Methods 12
    1.1.3 Manufacturing Processes for Fabrication of Coaxial
         Biofibres.. 21
  1.2 A Brief Overview of Fibre-Based Scaffolds 28
  1.3 Thesis Objectives 29
  References 30

2 General Experimental 47
  2.1 Components and Spinning Solutions 47
    2.1.1 Materials.. 47
    2.1.2 Gel Spinning Precursors........................... 47
    2.1.3 Graphene Oxide Liquid Crystal Dispersion 48
    2.1.4 PEDOT:PSS Dispersion 48
  2.2 Experimental Methods 49
    2.2.1 Spinning Techniques 49
  References 55

3 Preparation and Characterisation of Novel Hybrid Hydrogel
Fibres 57
  3.1 Introduction 57
  3.2 Experimental 59
    3.2.1 Materials.. 59
    3.2.2 Wet-Spinning of Chitosan, Alginate and Chit-Alg
         Coaxial Fibres................................... 59
    3.2.3 Characterisation Methods 60

  3.3 Results and Discussion 61
     3.3.1 Spinnability Versus Concentration 61
     3.3.2 Rheology 62
     3.3.3 Continuous Spinning of Coaxial Fibres 63
     3.3.4 Morphology of As-Prepared Fibres 63
     3.3.5 Mechanical Properties of As-Prepared Fibres 67
     3.3.6 Swelling Properties in SBF 68
     3.3.7 Thermogravimetric Analysis 69
     3.3.8 Cytocompatiblity Experiment 70
     3.3.9 In Vitro Release Measurement 71
  3.4 Conclusion 73
  References 73

4 Fabrication of Coaxial Wet-Spun Biofibres Containing
   Graphene Core 79
  4.1 Introduction 79
  4.2 Experimental 81
     4.2.1 Materials 81
     4.2.2 Fibre Spinning 81
     4.2.3 Characterisations of rGO and Coaxial Fibres 83
  4.3 Results and Discussion 85
     4.3.1 Optimization of Spinning Solutions 85
     4.3.2 Morphology of As-Prepared Fibres 88
     4.3.3 FTIR Spectroscopy Results 91
     4.3.4 Mechanical and Electrical Properties 92
     4.3.5 Surface Wettability of As-Spun Fibres 95
     4.3.6 Cyclic Voltammetry 95
     4.3.7 Raman Spectroscopy Results 98
     4.3.8 In Vitro Bioactivity Experiments 99
  4.4 Conclusion 101
  References 102

5 Development of One-Dimensional Triaxial Fibres as Potential
   Bio-battery Structures 107
  5.1 Introduction 107
  5.2 Experimental 109
     5.2.1 Materials 109
     5.2.2 Dispersion Preparation 109
     5.2.3 Coaxial Wet-Spinning of Chit-PEDOT
          and Alg-PEDOT 110
     5.2.4 Polymerisation of Pyrrole 110
     5.2.5 Fourier Transform Infrared 111
     5.2.6 Analysis 112

5.3 Results and Discussions 115
    5.3.1 Characterisation of Spinning Solutions 115
5.4 Conclusion... 131
References 133

6 Conclusion and Future Work 139
6.1 General Conclusion 139
6.2 Comparison of Fibre Properties 142
    6.2.1 Mechanical Properties 143
    6.2.2 Electrochemical Properties 143
6.3 Recommendations for Future Work 145

Abbreviations

Ag/Ag$^+$	Silver/silver ion reference electrode
Ag/AgCl	Silver/silver chloride reference electrode
aq.	Aqueous
°C	Degree of Celsius
Chit	Chitosan
CV	Cyclic voltammetry
EC	Electrochemical/Electrochemistry
EDS	Energy-dispersive X-ray Spectroscopy
ECM	Extracellular matrix
FepTS	Iron(III) p-toluene sulphonate
g	Gram
GO	Graphene oxide
HMW	high molecular weight
Hz	Hertz (frequency)
hr	Hour
i	Current
ISP	Isopropanol (Propan-2-ol)
L	Litre
LiCl	Lithium Chloride
M	Molar
mA	Milliampere
min	Minute(s)
mL	Millilitre
MMW	Medium molecular weight
NaCl	Sodium chloride
nm	Nanometre
OLED	Organic light-emitting diode
PCL	Polycaprolactone
PEDOT:PSS	Poly(3,4-ethylenedioxythiophene) polystyrene sulfonate
PEG	Polyethylene glycol

PLGA	Poly(lactic-co-glycolic acid)
PPy	Polypyrrole
PS	Polystyrene
Pt	Platinum
PTh	Polythiophene
RVC	Reticulated Vitreous Carbon
rpm	Revolutions per minute
s	Second(s)
S	Siemens
SEM	Scanning electron microscopy
t	Time
T	Temperature
TB	Toluidine Blue O
TBABF$_4$	Tetrabutylammonium bromide
TGA	Thermogravimetric analysis
UV-VIS	Ultraviolet-visible spectrophotometry
V	Potential
V$_d$	Drawing velocity
V$_i$	Injection rate (mL hr^{-1})
μ	Micro (prefix)
υ	Scan rate
γ	Shear rate (s^{-1})
η	Viscosity (Pa*s)
σ	Conductivity (Scm^{-1})

List of Figures

Fig. 1.1 Chemical structure of sodium alginate 3
Fig. 1.2 Chitosan chemical structures............................. 4
Fig. 1.3 2D honeycomb crystal lattice of sp^2 bonded carbon
          with a single atom thickness known as graphene 6
Fig. 1.4 Inherently conducting polymer structures represented
          in their undoped forms 8
Fig. 1.5 Polypyrrole polymerisation............................. 13
Fig. 1.6 Schematic of electrospinning [195]; Reproduced
          by permission of The Royal Society of Chemistry (RSC)
          on behalf of the Centre National de la Recherche
          Scientifique (CNRS) and the RSC 15
Fig. 1.7 Schematic of melt-spinning [195]; Reproduced by permission
          of The Royal Society of Chemistry (RSC) on behalf
          of the Centre National de la Recherche Scientifique (CNRS)
          and the RSC...................................... 18
Fig. 1.8 Schematic of a Lab scale wet-spinning line [195]; Reproduced
          by permission of The Royal Society of Chemistry (RSC)
          on behalf of the Centre National de la Recherche
          Scientifique (CNRS) and the RSC 19
Fig. 1.9 A schematic for coaxial electrospinning set up [279] 23
Fig. 1.10 Schematic for coaxial melt-electrospinning setup [291] 25
Fig. 1.11 Coaxial wet-spinning setup............................. 28
Fig. 2.1 A schematic of rotary wet-spinning method 49
Fig. 2.2 A schematic of coaxial spinneret 50
Fig. 2.3 a 25 mm × 10 mm pre-drilled brass block, b wet fibres
          inserted upright into the holes and protrude from the brass
          block to be imaged under LVSEM....................... 52

Fig. 2.4   Schematic of **a** lateral view and **b** cross-section of the triaxial
             spinneret   54
Fig. 3.1   Viscosities of spinning solutions of chitosan and sodium
             alginate   62
Fig. 3.2   The capability of producing an unlimited length of coaxial
             Chit-Alg (1) fibres as shown onto a collector   63
Fig. 3.3   The photographs of scaffold structure woven by coaxial
             fibres; imaged in **a** dry and **b** wet-state.   64
Fig. 3.4   Stereomicroscope images of the side view of wet **a** alginate,
             b chitosan, **c** coaxial Chit-Alg (1) and **d** dry Chit-Alg (1)
             fibre   65
Fig. 3.5   LV-SEM images of hydrated as-prepared **a** alginate, **b** chitosan
             and **c** Chit-Alg (1) cross section in SBF, **d** chitosan core
             arrangement in cross section **e** alginate sheath construction
             in the cross section   66
Fig. 3.6   Stress–strain curves obtained from tensile tests of alginate
             single and chitosan/alginate coaxial fibres using different
             $CaCl_2$ concentrations   67
Fig. 3.7   Swelling properties of coaxial wet-spun fibres in SBF
             as a function of the immersion time   69
Fig. 3.8   TG curves of alginate, chitosan and coaxial Chit-Alg (1)
             fibres.   70
Fig. 3.9   Coaxial chitosan-alginate fibres were analysed for their ability
             to support primary cell adhesion and growth. Murine (**a**) and
             human (**b**) myoblasts were observed to adhere and spread
             along the alginate surface of the fibres. **c** Calcein AM staining
             of human myoblasts revealed that the majority of cells
             remained viable for at least 7 days in culture and cells appeared
             to show some alignment with the surface features
             of the fibres   71
Fig. 3.10   Time dependent TB releasing behaviour of chitosan,
             alginate and Chit/Alg hydrogel fibres in SBF at 37 °C.
             Inset; burst release of coaxial fibres in the first 30 min   72
Fig. 4.1   A schematic of coaxial wet-spinning set up for producing
             Chit/GO coaxial fibres   82
Fig. 4.2   **a** Embedded fibres in Epoxy resin at room temperature,
             b fibre cross-sections through cut surface block   84
Fig. 4.3   The viscosity of spinning solution **a** chitosan 3% (w v^{-1}),
             b alginate 3% (w v^{-1}) and GO suspension (0.63 w v^{-1})
             as a function of shear rate.   87
Fig. 4.4   Illustrative optical microscopic images of **a** rGO fibre,
             b coaxial Chit/GO and **c** Alg/GO coaxial fibre in wet-state   88

Fig. 4.5   Representative LV-SEM micrographs of **a** hydrated rGO
            fibres, **b** higher magnification of rGO, **c** cross-section of
            coaxial Chit/GO fibre, **d** higher magnification of Chit/GO
            interface in coaxial fibres, **e** cross-section of coaxial Chit/GO
            fibre in compositional mode, **f** layered morphology of
            graphene core with dentate bends in wet-state, **g** cross-section
            of coaxial Alg/GO fibre and **h** higher magnification
            of Alg/GO interface in coaxial fibres   89
Fig. 4.6   Schematic representation of the functional groups
            on the edge of GO structure   90
Fig. 4.7   Cross sections of **a** rGO fibre, **b** higher magnification
            of rGO fibre, **c** coaxial Chit/GO fibre, **d** surface of Chit/GO
            oxide fibre and **e** tied coaxial fibre and **f** brittle rGO fibre   92
Fig. 4.8   FTIR spectra of chitosan and Chit/GO showing
            the possible chemical reactions between those   93
Fig. 4.9   Stress-strain curves obtained from the uniaxial tensile test
            on chitosan and coaxial Chit/GO fibres. Inset; Stress-strain
            curves obtained from the uniaxial tensile test on rGO fibres. . . .   94
Fig. 4.10   Water contact angle measurement on suspended fibres.   95
Fig. 4.11   Cyclic voltammograms of **a** rGO and **b** Chit/GO fibres;
            potential was scanned between −0.8 V and +0.8
            (*vs.* Ag/AgCl) in PBS electrolyte and 1 M aqueous NaCl
            solution at 50 mVs^{-1}   96
Fig. 4.12   The Cyclic voltammograms of Chit/GO oxide fibres before
            and after 40 cycles in (**a**) PBS electrolyte and (**b**) 1 M aqueous
            NaCl solution at 50 mVs^{-1}; potential was scanned between
            −0.4 V and +0.3 (*vs.* Ag/AgCl)   97
Fig. 4.13   Raman spectra from Chit/GO, reduced Chit/GO and rGO
            fibres.   99
Fig. 4.14   Cell viability (**a**) and cell density (**b**) quantified from images
            of live/dead stained L-929 cells grown on coaxial (**c**) and rGO
            (**d**) fibres and cell population growing on the underlying
            microscope slide. Values in **a** and **b** obtained by image
            analysis represent the average of at least 20 images
            (900 × 900 μm each), or at least 50 mm of fibre length,
            and error bars show one standard deviation of the mean.
            Scale bars in **c** and **d** represent 200 μm   100
Fig. 4.15   Confocal microscopy image showing cell attachment
            on the surface of coaxial fibre along its length   101
Fig. 4.16   Differentiated neural cell line (PC-12 cells) on **a** coaxial
            and **b** reduced graphene oxide fibres, showing small degree
            of differentiation (neurites indicated by arrows)   101
Fig. 5.1   The electrostatic absorption between PEDOT: PSS
            and chitosan   111

Fig. 5.2   Hydrogel sample holder containing 5 channels varying
            in length between RVC electrodes from 0.5 to 2.5 cm
            (height, 6 mm; channel width, 5 mm)..................... 113
Fig. 5.3   Viscosities of spinning solutions of PEDOT:PSS with
            and without PEG, chitosan and sodium alginate solutions 116
Fig. 5.4   Electrical impedance analysis of chitosan hydrogel 3%
            (w v^{-1}); **a** Bode plot and **b** Impedance as a function
            of sample length in wet-state fibres 118
Fig. 5.5   Viscosity changes of **a** Alg and **b** chitosan spinning
            solutions with and without addition of NaCl 1% (w v^{-1})...... 120
Fig. 5.6   Stereomicroscope images of (**a–c**) surface of wet Chit-PEDOT
            (+PEG) fibres without post-treatment 121
Fig. 5.7   Stereomicroscope images of surface and **b** cross-sections
            of (**a**), **b** Chit-PEDOT (+PEG) and **c** and **d** Alg-PEDOT
            (+PEG), respectively................................. 121
Fig. 5.8   LV-SEM images of hydrated cross sections of as-prepared
            a Chit-PEDOT, **b** Chit-PEDOT (+PEG) and **c** higher
            magnification of Chit-PEDOT (+PEG).................... 122
Fig. 5.9   LV-SEM images of hydrated cross sections of as-prepared
            a Alg-PEDOT, **b** Alg-PEDOT (+PEG), **c** higher
            magnification of Alg-PEDOT (+PEG) and **d** re-hydrated
            Alg-PEDOT (+PEG) fibres............................. 122
Fig. 5.10   SEM images of cross sections of as-prepared coaxial fibres
            of **a** Chit-PEDOT (+PEG), **b** Alg-PEDOT(+PEG) and higher
            magnifications of **c** Chit-PEDOT (+PEG) and **d** Alg-PEDOT
            (+PEG) fibre, surface pattern of **e** Chit-PEDOT (+PEG)
            and **f** Alg-PEDOT (+PEG) fibre, **g** cross section, **h** higher
            magnification of cross-section and **i** surface of
            PPy-Chit-PEDOT (+PEG) triaxial fibres.................. 124
Fig. 5.11   EDS maps from cross-sections of as-prepared **a** Chit-PEDOT
            (+PEG), **b** Alg-PEDOT (+PEG) and **c** PPy-Chit-PEDOT
            (+PEG) fibres; where **a**, **b** and **c** (1) (dark blue), **a** and **c** (2)
            (red) and **b** (2) (light blue) shows the elemental maps
            of sulphur (S), carbon (C) and calcium (Ca), respectively 125
Fig. 5.12   Mechanism of how antioxidants reduce free radicals
            by giving the free radical an electron which inhibits other
            oxidation reactions 126
Fig. 5.13   An oxidation reaction of alginate upon addition
            of the oxidant...................................... 126
Fig. 5.14   FTIR spectra of alginate and alginate-Fe*p*TS 127
Fig. 5.15   Cyclic voltammograms of PEDOT:PSS (+PEG), Chit-PEDOT
            (+PEG) and Alg-PEDOT (+PEG) fibres in **a** PBS solution and
            b 0.1 M TBABF$_4$; potential was scanned between −0.8 V and
            +0.8 at 10 mV s^{-1} 130

Fig. 5.16   Primary cell attachment and proliferation on fibres after 5 days
             of differentiation. **a** Alg-PEDOT (+PEG) and Chit-PEDOT
             (+PEG) fibres from **b** bottom **c** side and **d** top view,
             respectively. Scale bars represent 100 μm.   132
Fig. 6.1   Mechanical properties of all fibre types produced
             in the course of this study   144
Fig. 6.2   Electrochemical properties of all fibre types produced
             in the course of this study   145

List of Tables

Table 1.1   Some of the representative electrospun systems studied
for biomedical applications 17
Table 1.2   Some of the representative wet-spun systems studied
for biomedical applications 20
Table 3.1   Thickness of sheath and core, μm, in wet-state as a function
of wet-spinning condition............................. 65
Table 3.2   Comparison of mechanical properties of solid and coaxial
biofibres.. 68
Table 4.1   Mechanical properties and conductivity results of GO, rGO,
chitosan and Chit/GO fibres........................... 93
Table 4.2 I_D/I_G as observed in the Raman spectra for Chit/GO,
reduced Chit/GO and rGO fibres 98
Table 5.1   Ionic conductivity results of hydrogels using two kinds
of salts... 119
Table 5.2   Mechanical properties of coaxial and triaxial fibres 128

Chapter 1
Introduction and Literature Review

1.1 Introduction

1.1.1 Material Considerations for Biomedical Applications

For an ideal scaffolding material, properties are required that include biocompatibility, suitable microstructure, desired mechanical strength and degradation rate as well as most importantly the ability to support cell residence and allow retention of metabolic functions. Numerous strategies currently used to engineer tissues depend on employing a material scaffold. These scaffolds serve as a synthetic extracellular matrix (ECM) to organize cells into a 3D architecture and to present stimuli, which direct the growth and formation of the desired tissue. Depending on the tissue of interest and the specific application, the required scaffold material and its properties will be quite different. A wide range of materials is known to be utilised as cell-supporting materials in biomedical applications including natural and synthetic polymers [1–3], metals [4], ceramics [5, 6] and alloys [7, 8]. Aside from the specific materials used in certain applications such as orthopaedics, dental implants as well as artificial vascular materials, the focus of this thesis is on the role of naturally occurring hydrogels and organic conductors to develop biofibres with the final use as biocompatible electrodes for the purpose of signal recording or electrical stimulation or the building blocks of tissue scaffolds.

Herein, materials are categorised into ionically conducting and electrically conducting materials. Non-conducting materials have shown a great promise to be used as templates for tissue engineering and implantable devices. Conducting materials have also been used as templates for cell attachment and growth while providing a pathway for electrical stimulation of cells. Hydrogels including alginate and chitosan as well as conducting graphene, PEDOT:PSS and PPy are introduced and explained as follows.

© Springer Nature Switzerland AG 2018

A. Mirabedini, *Developing Novel Spinning Methods to Fabricate Continuous Multifunctional Fibres for Bioapplications*, Springer Theses, https://doi.org/10.1007/978-3-319-95378-6_1

1.1.1.1   Natural Hydrogels

Natural polymers can be considered as the most undeniable biomaterials used clinically [9] in different shapes. Polymers of natural origin are attractive options, mainly due to their similarities with ECM as well as their chemical versatility and biological performance [3]. A variety of hydrogels, three-dimensional covalently crosslinked polymer networks with a high number of hydrophilic groups, capable of accommodating large amounts of water [10, 11], are being employed as scaffold materials. They are composed of hydrophilic polymer chains, which are either synthetic or natural in origin. The structural integrity of hydrogels depends on crosslinks formed between polymer chains via various chemical bonding and physical interactions which make them to be resistant to be solubilized. Hydrogels used in bioapplications are typically degradable, can be processed under relatively mild conditions, have mechanical and structural properties similar to many tissues and the ECM, and can be delivered in a minimally invasive manner.

Hydrogels demonstrated a distinct efficacy as matrices for 3D cell culture since they are very similar to living tissues and ECM, such as a soft and rubbery yet deformable natures and low interfacial tension with biological fluids [12]. In addition, they can use biologically relevant electrolytes makes them well suited for applications within biology [13, 14]. Another unique characteristic of biomimetic hydrogels is that they undergo huge volume changes, which occur in relatively narrow ranges of changes of temperature, pH, and ionic strength [13, 15].

Polysaccharides are a typical group of natural biopolymers showing great swellability that makes them ideal candidates for making hydrogels. Polysaccharides are high molecular weight polymeric carbohydrates formed of repeating monosaccharide units [16]. Polysaccharides are advantageous for biomedical applications due to their wide availability, low cost as well as the presence of functional groups in the polymer chain [17]. They offer a wide diversity in structure and properties due to their wide range of molecular weight and chemical composition. Alginate [18–21] and chitosan [18, 22, 23] are considered as the most extensively used gel-forming polysaccharides [20] for cell growth from natural sources. They were chosen and used in this study mainly due to their several unique properties including biodegradability, biocompatibility, low toxicity, promoting attachment, migration, proliferation and differentiation of cells and antimicrobial activity [24] as well as ease of fabrication and availability.

1.1.1.1.1   Sodium Alginate

The anionic polymer alginic acid or alginate is a natural polysaccharide obtained brown seaweeds. Since it was discovered by Stanford [25] in 1881, alginate has been used in a wide range of industries, such as food, textile printing, paper and pharmaceuticals, and for many other novel end-uses. Alginate is a linear, binary copolymer composed of 1, 4-linked β-D-mannuronic acid (M) and α-L-guluronic acid (G) monomers. Alginates are extracted from algae using a basic solution [20].

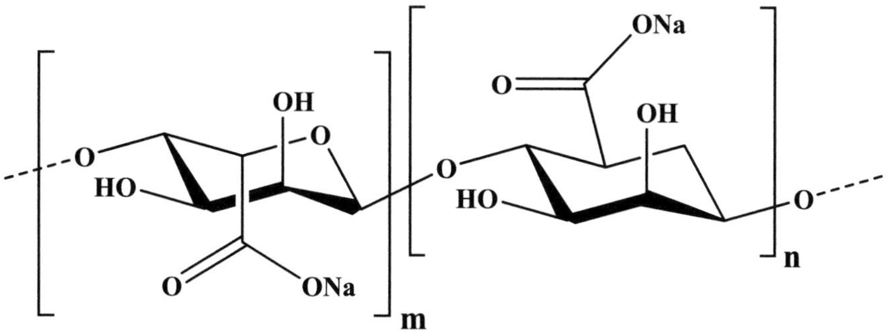

Fig. 1.1 Chemical structure of sodium alginate

The extracted material is then reacted with acid to form alginic acid. The composition of alginate (the ratio of the two uronic acids and their sequential arrangements) varies with the source. Salts of alginic acid with monovalent cations such as sodium alginate are all soluble in water [26] capable of holding a large amount of water. Alginate has been extensively used as a scaffold for liver [27], bone [19], nerve [28] and cartilage engineering [29]. Even though, alginates are non-toxic and biocompatible, using them for biomedical applications has several drawbacks. Alginates are mechanically very weak in wet condition, therefore it should be blended or modified or copolymerized with other biopolymers before being used as a structural scaffold. More importantly, it shows poor cellular adhesion. The chemical structure of sodium alginate is demonstrated in Fig. 1.1.

By forming alginate into fibres, novel biomaterials are attainable which can be processed further into woven, non-woven, braided, knitted and many other kinds of composite structures. In the wet-spinning process (as explained in detail in Sect. 1.3.2.3) in which alginate is transformed from powder into a fibrillar-shape, alginate powder is needed to be dissolved in water and stirred properly to form a homogenous solution first. The final properties of wet-spun alginate fibres highly depend on a number of factors, such as chemical structure and molecular weight of the alginate, composition of the coagulation bath, drawing ratio, temperature and feeding rates, etc. The spinning solution is one of the first main considerations in the wet-spinning process which determines the production efficiency. The fibre final performances strongly depend on several parameters including concentration, temperature and pH of the spinning solution [30]. A concentrated sodium alginate solution can be extruded through spinneret holes into a calcium chloride ($CaCl_2$) bath, whereby the high acid content allows alginic acid to undergo spontaneous and mild gelling in the presence of di- or trivalent cations [31]. Thus, it is possible to use a variety of metal ions such as zinc [32], silver [33–35] or other bioactive metal ions to precipitate sodium alginate solution during the wet-spinning process as tried previously, too. Among divalent ions, calcium has found greatest popularity for gel formation of alginate fibres mainly because its salts are cheap, readily accessible and cytocompatible [30]. Since processing takes place in an aqueous solution and in an aqueous coagulation

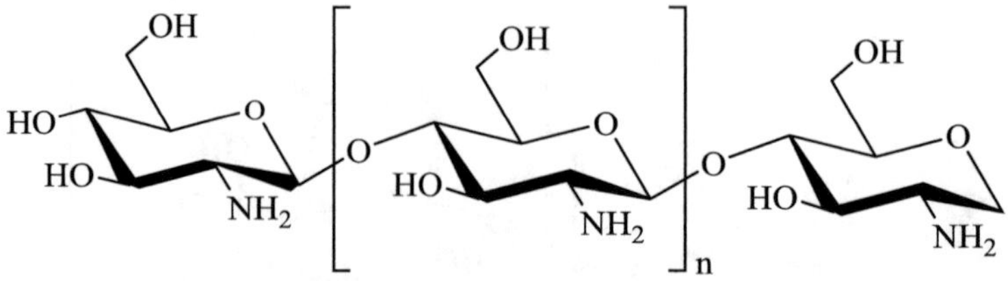

Fig. 1.2 Chitosan chemical structures

bath at a neutral pH, many bioactive materials, such as drugs and enzymes, can be combined into the alginate fibres, without loss of their bioactivity. On the other hand, calcium alginate fibres have proven to be unstable structures as tissue scaffolds or drug vehicles for in vivo usages [36, 37].

1.1.1.1.2 Chitosan

Chitosan is a semi-crystalline natural polysaccharides with a totally different nature with that of alginate which has recently generated great interest for its potential in clinical and biological applications [38–41] such as artificial skin, tissue engineering and controlled drug delivery [26, 42]. The cationic polymer chitosan originates from crustacean skeletons [38]. Structurally, chitosan is a semi-synthetically derived aminopolysaccharide which is the N-deacetylated product of chitin, i.e. poly-$(1 \rightarrow 4)$-2-amino-2-deoxy-β-D-glucose [2, 9, 40, 43]. Chitosan shows an enhanced hydrophilicity compared to that of chitin which results in a considerable loss of tensile strength in wet state [44]. This hydrogel is highly reactive due to free amine groups and is readily soluble in weakly acidic solutions resulting in the formation of a cationic polymer of chitosan acetate with a high charge density. These solutions generally have high solution viscosities due to the phenomenon known as the polyelectrolyte effect [39, 45–47]. Porous chitosan matrix has been used as a scaffold for skin [48–50], liver [51], bone and cartilage [52–55], cardiac [56–58], corneal [59] and vascular regenerative tissue remodelling [22, 60, 61], It has also been applied in controlled drug delivery is different shapes such as spheres, films or fibres [62, 63]. The chemical structure of chitosan is shown in Fig. 1.2.

Chitosan can be produced in a variety of forms including films, fibres, nanoparticles and microspheres. There have been many attempts by several groups into aqueous basic coagulating baths to produce chitosan fibre [24, 43, 64, 65]. For the purpose of wet-spinning (detailed in Sect. 1.1.2.3), the chitosan solution is generally extruded into an alkaline solution such as aqueous NaOH as the coagulation bath which forms the fibres. The coagulation rate, which also includes the regeneration of the free amine form of chitosan, is also expected to be influenced by high solution viscosity. Nevertheless, the strong alkaline condition (pH > 12) needed to form chitosan-based structures, can limit its utilization for loading most of the drugs or bioactive molecules into it.

1.1.1.2 Conducting Materials

A wide range of biomaterials has been already used in developing structures for biomedical applications. However, it is hard to find a biomaterial that gathers all the requirements for specific biomedical applications. Conducting materials are typically utilised as a connector between cells and electrical devices to pass or receive electrical signals to and from cells. Electrical stimulation has been able to provide beneficial effects for regeneration of tissue: muscle [66–69], nerve [70, 71] and bone [72]. Moreover, they could improve the mechanical properties of biomaterials. A range of metallic electrodes, most commonly Pt or titanium based, have conventionally been used as the preferred metal used for electrodes [69, 73]. However, metals are mechanically significantly stiffer than the neural tissue with which it interfaces [74]. For example, the elastic modulus of Pt is measured to be about 164 GPa [75], but most neural tissue has a modulus of less than 100 kPa [76]. Using organic conductors in particular CPs, have been shown to impart a softer electrode interface, around 1 MPa [77]. Therefore, it is expected that these materials can be used to dampen or mediate the mechanical difference between a metal electrode and the tissue with which it interfaces. It has also been shown that to stimulate light precepts by electrical stimulation of the retina, a charge density between 48 and 357 μC/cm^2 is required [78], but the electrochemical injection limit of metals electrodes such as Pt has been reported as ranging from 20 to 150 μC/cm^2 [74]. This small range of overlap means that metal electrodes cannot be safely reduced in size and still maintain safe charge injection at a therapeutic level. More recently the use of organic conductors has attracted attention since they can be loaded with bioactivity enhancing the electro-cellular communication process [79, 80]. Polymers have been traditionally considered to be electrical insulators. Recently, the progress made in chemical synthesis of organic conductors has brought a rich variety of conducting organic materials. These organic conductors can be categorized into inherently conducting polymers and carbon-based materials including carbon nanotubes and graphene. Use of organic conductors including CPs has shown that tailored approaches can be used to create multi-functional electrode arrays which not only improve the electrode material properties but also provide biomolecules to aid in the establishment of a chronically stable neural interface. The background, chemical structures, synthesis methods and EC properties of the conducting polymers used in this work, specifically PEDOT:PSS, PPy and graphene, are described in the following sections.

1.1.1.2.1 Graphene

Carbonaceous materials such as CNTs, graphite, fullerene, graphene and graphene oxide have recently attracted the attention of many researchers as conducting pathways incorporated into biomaterials. Graphene is a carbon-based material which is a layer of tightly packed two-dimensional (2D) honeycomb crystal lattice of sp^2 bonded carbon with a single atom thickness [81] as shown in Fig. 1.3. Graphene has shown extraordinary optical properties, thermal conductivity and outstanding

Fig. 1.3 2D honeycomb crystal lattice of sp^2 bonded carbon with a single atom thickness known as graphene

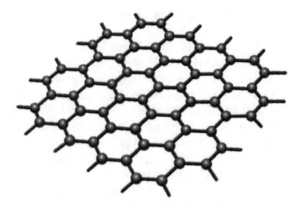

mechanical properties (Young's modulus up to ~1TPa). One of the most efficient methods for large scale and low cost production of graphene derivative is oxidative exfoliation of natural graphite followed by a chemical reduction. Using this method, most of the functional groups such as hydroxide, carboxyl, epoxide and carbonyl exist in the chemical structure of GO will be removed. These functional groups make graphene oxide (GO) hydrophilic and stable in an aqueous medium. Therefore, the reduction process produces highly conductive chemically converted graphene (CCG) or reduced graphene oxide (rGO) which is hydrophobic and insoluble in aqueous media [82]. Low solubility may cause irreversible aggregation which limits further processing. Li et al. have shown that stable dispersions of rGO is possible to be achieved without using by controlling the pH during the reduction process [83]. Yet, this dispersion is not directly spinnable by itself. Addition of spinnable polymers to the dispersion provides this opportunity to form it into a fibre. Then again, to achieve considerable electrical conductivity, exceeding the percolation threshold, higher loadings (typically 10–80%) are required. However, rGO dispersions are usually obtained and maintained at low concentrations.

The recent success in assembling graphene sheets into macroscopic fibres has inspired extensive interest in these materials because of the lower cost of graphene fibres (GFs) compared with CNTs and commercial carbon fibres, and their practical importance for specific applications [84]. LC GO structure allows for the dispersion of GO at high enough concentrations suitable for efficient alignment and effective coagulation. Gao et al. might be the first who reported of fabrication of GO fibres *via* a wet-spinning approach by loading the GO dispersions into glass syringes and injecting them into a coagulation bath of a 5 wt% NaOH/methanol solution [85]. Spinning of liquid crystalline (LC) suspensions of large sheet graphene oxide (GO) in water has been recently reported by several research groups [85–87]. Use of large GO sheets has enabled the use of a wet-spinning route to produce strong fibres by extruding them through a thin nozzle into an appropriate coagulation bath which can be easily converted to electrically conducting graphene fibres by using an appropriate chemical reducing agent. Preparation of GO fibres using various coagulation baths such as $CaCl_2$, acetone, NaOH as well as chitosan has been reported earlier [87]. In addition, it is suggested that graphene oxide could undergo quick deoxygenation when exposed to strong alkali solutions at moderate temperatures which was also previously studied [88, 89].

1.1.1.2.2 Inherently Conducting Polymers

Inherently conducting polymers (ICPs) were discovered in 1977 with the 10^9 times increase in electrical conductivity (σ) of polyacetylene (PAc) through halogen doping to as high as 10^5 Scm^{-1} by means of chemical modification [90] (in this case leading to partial oxidation or reduction by reaction with electron acceptors and donors known as a doping process [91]. These organic polymers that possess the electrical, electronic and optical properties of a metal while retaining the mechanical properties, processability, etc. commonly associated with a conventional polymer, is termed an ICP or more commonly as a "synthetic metal" [92]. To date a tremendous amount of research has been carried out in the field of conducting polymers, while the broader significance of the field was recognised in the year 2000 with the awarding of the Nobel Prize for Chemistry to the three discoverers of ICPs, Shirakawa, MacDiarmid and Heeger [92, 93]. Since the discovery of conducting PAc, a number of additional ICPs have been developed, including polypyrrole (PPy) [94–98], polyaniline (PAni) [99–101], polythiophene (PTh) [102, 103], poly(p-phenylenevinylene) (PPV) [104, 105], poly(3,4-ethylene dioxythiophene) (PEDOT) [92, 106–108], and polyfuran (PF) [109]. Subsequently, extensive characterisation of ICPs including electrical, thermal and environmental stability, as well as their processability, displayed their high potentials to be used in organics, renewable energy as well as biomedical applications. The structures of selected conducting polymers developed over the last 30 years are illustrated in Fig. 1.4. The most significant conducting polymers with regard to technological fibres are PAni, PPy, PTh and PEDOT.

ICPs could be synthesised either by chemical or electrochemical methods, with chemical synthesis being the preferred option when large quantities of polymer are required. This procedure involves the addition of a strong oxidising agent such as $FeCl_3$ followed by deposition of the monomer to initiate the free radical polymerisation [110]. EC synthesis is usually performed in a 3-electrode cell comprising a working electrode, an auxiliary electrode and a reference electrode, whereby an insoluble ICP film forms on the working electrode initiated by applying either a certain amount of voltage or current. Among the synthesis methods, EC synthesis in particular presents several distinct advantages such as the absence of catalyst, direct grafting of the doped conducting polymer onto the electrode surface, easy control of film thickness by controlling the deposition charge, and the possibility to perform in situ characterisation of the polymerisation process by EC and/or spectroscopic techniques.

The electrical, electrochemical and physical properties of ICPs could be tailored for particular final applications. Not surprisingly therefore they have already been applied in a wide variety of areas such as biosensors [111, 112], biomedical

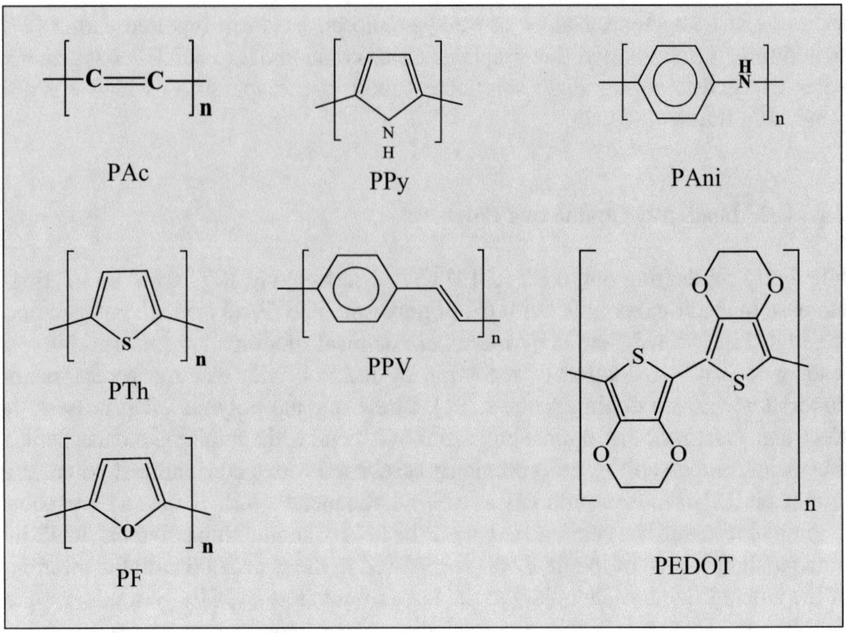

Fig. 1.4 Inherently conducting polymer structures represented in their undoped forms

applications [113–115], electrical stimulation of cells [116, 117] and drug delivery [118]. It has also been investigated as a material for "artificial muscles" that would offer numerous advantages over traditional motor actuation [119, 120]. Conducting polymers such as PANi or PPy are not biodegradable. However, they are being used in combination with biodegradable polymers such as polylactic-co-glycolic acid (PLGA), polylactic acid (PLA) or polyvinyl alcohol (PVA) was required. PEDOT is a well-studied intrinsically conducting polymer that is rendered solution-processable when doped with acidic PSS [121]. The processability of PEDOT:PSS has naturally meant that relatively few studies have considered PEDOT:PSS within composite fibres. Nevertheless, composite PEDOT:PSS fibres are at the center of attention due to their high conductivity and multiple applications such as sensors and drug delivery [112, 122, 123]. PPy is another most preferred conducting polymer for biomedical applications since it can be synthesized in a natural pH condition. Moreover, it has been shown to be an appropriate substrate supporting cell attachment and growth [69, 124].

Printing and fibre spinning technologies can be regarded as two of the most prominent techniques investigated for the development of devices based on ranges of materials including ICPs and graphene. Printing is a fast, old and inexpensive method that is used for mass fabrication of advanced conducting components [125]. In recent years, increasing efforts have been focused on the printing of conducting polymer-based devices [126]. Printing is a reproduction process in which ink is applied to a substrate in order to transmit information such as images, graphics and text. Printed materials must form a solid, continuous conducting film following solvent removal. The solvent plays significant roles such as compatibility with the conducting polymer, stability in solution and appropriate rheological and surface energy characteristics. Printing technologies that require a printing plate are known as conventional methods and include lithography (offset), gravure, letterpress and screen-printing. Non-impact printing (NIP), such as inkjet printing or electrophotography, uses laser technology and does not require a printing plate [127]. Printing provides a convenient route to the deposition of conducting polymers with spatial resolution in the x, y plane in the order of tens of microns and makes layer thicknesses in the order of 100 nm feasible. The birth of 3D-printing goes back to 1984 when as Charles Hull invented stereolithography which enabled a tangible 3D object to be created from a 3D model [128]. Varieties of conducting polymers have been processed earlier to become printable including PANi [129, 130], PPy [131, 132], and PTh [133].

Extensive advances have been also made during the last three decades in the fundamental understanding of fibre spinning using conducting polymers. Conducting polymers must undergo processing steps in order to attain the desired form [125]. The very first attempts to achieve optimal conditions for the spinning of fibres from PAni were begun in the late 1980s [134–136]. A few years later Mattes et al. pioneered the processing of PAni into fibre form through a dry-wet-spinning process [84, 137]. PEDOT:PSS and PPy are two of the most investigated conducting polymers which were also used in this research. In the following sections it is attempted to provide an overview on the used conducting polymers.

1.1.1.2.3   Poly(3,4-Ethylene Dioxythiophene)

Amongst the wide variety of conducting polymers, those derived from thiophene and its derivatives show good stability toward oxygen and moisture in both doped and neutral states [138]. This combined with favourable electrical and optical properties has led to the application of PThs in electrochromic displays, protection of semiconductors against photocorrosion, and energy storage systems [139]. PTh results from the polymerisation of thiophene, a sulfur heterocycle, which may be rendered conducting when electrons are added or removed from the conjugated π-orbitals *via* doping. PThs have been prepared since the 1980s by means of two main routes, namely chemical, and cathodic or anodic EC synthesis [140]. In the latter half the 1980s, scientists at the Bayer AG research laboratories developed the PTh derivative PEDOT (or PEDT), which was initially developed with the aim of providing a soluble conducting polymer [141]. 3,4-ethylene dioxythiophene (EDOT) polymerises effectively, leading to PEDOT films that adhere well to typical electrode materials. PEDOT benefits from the absence of undesirable α,β- and β,β-couplings between monomer units, while its electron-rich nature plays a significant role in the optical, EC, and electrical properties of subsequent polymers based around the PEDOT building block [142]. PEDOT is characterised by stability, high electrical conductivity (σ) (up to 1000 Scm^{-1}), moderate band gap, low redox potential, and transparency in the oxidised state [141]. Initially PEDOT was found to be insoluble in common solvents. However, this was successfully overcome by using poly(styrenesulfonic acid) (PSS) as the dopant during its chemical synthesis. The resulting stable dark-blue aqueous dispersion of PEDOT:PSS is now commercially available and applied in antistatic coatings [143], electrode materials [144], organic electronics [145], transparent electrodes, capacitors [146], touchscreens, organic light-emitting diodes (OLED), microelectrodes and sensors [142, 147].

Several fabrication methods have been employed so far for producing fibres from PEDOT derivatives such as electrochemically synthesis [147], nanofibre seeding method [142], chemical polymerisation without employing a template [148], Preparation of microfibres from PEDOT:PSS was first reported by Okuzaki and Ishihara for the first time *via* wet-spinning into an acetone bath as the coagulant [149] where the effects of spinning conditions on fibre diameter (which ranged between 180 and 410 μm), electrical conductivity, microstructure and mechanical properties were investigated. Shortly thereafter, Okuzaki et al. fabricated highly conducting PEDOT:PSS microfibres with 5 μm diameter and up to 467 Scm^{-1} electrical conductivity by wet-spinning followed by ethylene glycol post-treatment [150]. Dipping in ethylene glycol (two-step wet-spinning process) resulted in a 2–6 fold increase in electrical conductivity from 195 to 467 Scm^{-1} and a 25% increase in tensile strength after drying from 94 to 130 MPa. Characterisation with X-ray photoelectron spectroscopy, X-ray diffractometry and atomic force microscopy led to the conclusion that the removal of insulating PSS from PEDOT:PSS grain surfaces and crystallization were responsible for the enhanced electrical and mechanical properties of the microfibres. This work opened a new way for scientists to prepare relatively long PEDOT:PSS fibres using a straightforward method. Jalili et al. simplified the method to a one-step process to prepare microfibres by employing a wet-spinning formulation consisting of an aqueous blend of PEDOT:PSS and poly(ethylene glycol), where the need for post-spinning treatment with ethylene glycol was eliminated and fairly high electrical conductivities of up to 264 Scm^{-1} were achieved [151].

1.1.1.2.4 Polypyrrole

Amongst the conducting polymers, PPy and its derivatives are of particular interest owing to rather straightforward synthetic procedures, reasonable stabilities in oxidised states in air and solvents, low cost, biocompatibility and availability of monomer precursors [152, 153]. However, it was not until 1977 that PPy attracted significant attention [92]. Dall'Olio et al. published the first report of the synthesis of a PPy film, which exhibited 8 Scm^{-1} electrical conductivity, by electrolysis of a pyrrole solution in the presence of sulphuric acid in 1968 [154]. The major breakthrough with regard to the routine synthesis of PPy, however, was achieved by Diaz et al. when they reported a highly conducting (100 Scm^{-1}), stable and flexible PPy film prepared by electrolysis of an aqueous solution of pyrrole [94]. Chemical methods in addition to EC methods have also been employed for the synthesis of PPy, such as photochemistry, metathesis, concentrated emulsion, inclusion, solid-state, plasma, pyrolysis and soluble precursor polymer preparation [153]. Nevertheless, it should be taken into account that EC polymerisation provides a number of advantages over chemical methods, such as the final form of reaction product (an electroactive film attached to the electrode surface), high electrical conductivity, and control over film mass, thickness and properties [153].

PPy demonstrates high electrical conductivity, good EC properties, strong adhesion to substrates and thermal stability [155, 156]. The heteroatomic and extended π-conjugated backbone structure of PPy provides it with chemical stability and electrical conductivity, respectively [154, 157]. PPy exhibits a wide range of surface electrical conductivities (10^{-3} Scm^{-1} $< \sigma <$ 100 Scm^{-1}) depending on the functionality and substitution pattern of the monomer and the nature of the counterion or dopant [158]. Not surprisingly therefore PPy has already been applied in a wide variety of areas such as rechargeable lithium batteries [159, 160], medical applications [114, 115] and drug delivery [122, 161]. It has also been investigated as a material for "artificial muscles" that would offer numerous advantages over traditional motor actuation [119, 120]. However, PPy usually takes the form of an intractable powder following chemical polymerisation [162] which is insoluble in most organic solvents [163, 164]. These characteristics may be largely attributed to the presence of strong interchain interactions and a rigid structure. The low water solubility and poor processability of PPy mean that there are few reports of pristine PPy fibres [165]. It follows that PPy may be considered as the most utilised conducting polymer in making composite fibres. Over the past two decades, a variety of materials including polymers sheets, glass, polymer and inorganic particles, clays, zeolites, porous membranes, fibres and textiles, and soluble matrices have been demonstrated as appealing substrates for PPy. Due to the good adhesion force between PPy and various substrates [166], conducting composites may be prepared that retain the inherent properties of both PPy and the substrate [167]. These substrates include carbon, graphite [168], glass [169], and polymeric fibres [170, 171]. In general, the conductivity of PPy/fibre composites is directly related to PPy loading, ratio of oxidant to dopant, and fibre structure [172].

A number of researchers attempted to improve polymer solubility involving alkyl group substitution at the 3- and 4-positions or at the nitrogen atom of the pyrrole ring [162]. Another technique that has proven successful has been the use of long chain surfactant dopants such as sodium dodecyl benzene sulfonate (DDS) [173, 174], di(2-ethylhexyl) sulfosuccinate sodium salt (DEHS) [175], and polystyrene sulfonate [176]. PPy doped with such surfactants were soluble in a number of solvents including *m*-cresol, NMP, dimethyl sulfoxide (DMSO), dimethyl formamide (DMF) and tetrahydrofuran (THF) [162]. Few reports exist that consider the wet-spinning of soluble PPy into continuous fibres, despite initial attempts [177]. This question was essentially abandoned for a number of years until Foroughi et al. published the first report on the production of continuous conducting PPy fibres through wet-spinning [163], which showed electrical conductivity of ∼3 Scm^{-1} and elastic modulus of ∼1.5 GPa. Although a number of researchers continue to seek new methods to produce wet-spun PPy fibres, no additional reports have been published.

CVD (also known as vapour phase polymerisation) is a straightforward and rapid method to deposit PPy onto various substrates, and has been used widely to produce composite PPy fibres [124, 166, 171, 178–181]. Although this method has the advantage of simplicity, the highest reported electrical conductivity of fibres prepared this way was only 0.68 Scm^{-1} [181], likely due to the formation of only a thin layer of conducting PPy. Nair and co-workers were the first to merge electrospinning with CVD for the synthesis of electrically conducting composite PPy nanofibres [171]. This approach provided the advantages of electrospinning while at the same time circumventing the intractability of PPy. Figure 1.5 shows the polymerisation of polypyrrole from a bulk solution containing pyrrole monomer. The electrodeposition proceeds by adsorption of a monomer unit onto a surface of the working electrode, to form a pyrrole cation radical through the oxidation process. These cations combine together or with neutral monomers present in the solution, to form a dimer, which undergoes double deprotonation to provide a neutral molecule. Comparing to monomer units, dimer radicals are more stable showing a lower oxidation potential. The polymer chain growth then occurs through favoured coupling between monomers and dimers. Formation of a π-conjugated backbone and its heteroaromatic structure brings about electrical conductivity as well as chemical stability. To maintain the charge balance along the polymer backbone, the positive charges made during the polymerisation are necessary to be combined with negatively charged counter ions known as a doping process [125].

Chronakis et al. reported for the first time a method to prepare nanofibres using a mixture of PPy and PEO [182]. In 2007, a microfluidic approach was described by others for fabricating hollow and core/sheath PPy nanofibres by electrospinning [183]. The benefits of using microfluidic devices for nanofibre synthesis include rapid prototyping, ease of fabrication, and the ability to spin multiple fibres in parallel through arrays of individual microchannels. PPy composite core–shell nanostructures were also successfully prepared using PAN, PS and Polyamide 6 (PA6) solutions [184]. It is worth noting that a large number of prepared PPy composite fibres have been employed for sensor applications [166, 185, 186].

1.1.2   Fabrication Methods

The fabrication methods for developing three dimensional structures to be utilised for biological applications have risen due to the inability of two dimensional structures to mimic the extracellular matrix accurately. To design a three dimensional architecture which imitates the ECM, several parameters such as geometry, mechanical and surface properties as well as biocompatibility are required to be taken into account [187]. Formation of tubular electroactive structures has recently attracted a great deal of attention as mentioned previously due to several unique advantages they offer to compare to the simple structures. Formation of coaxial fibres, wherein the organic conductor is encapsulated within a more cytofriendly material, could provide an alternate fabrication option which enables both improved electrical stim-

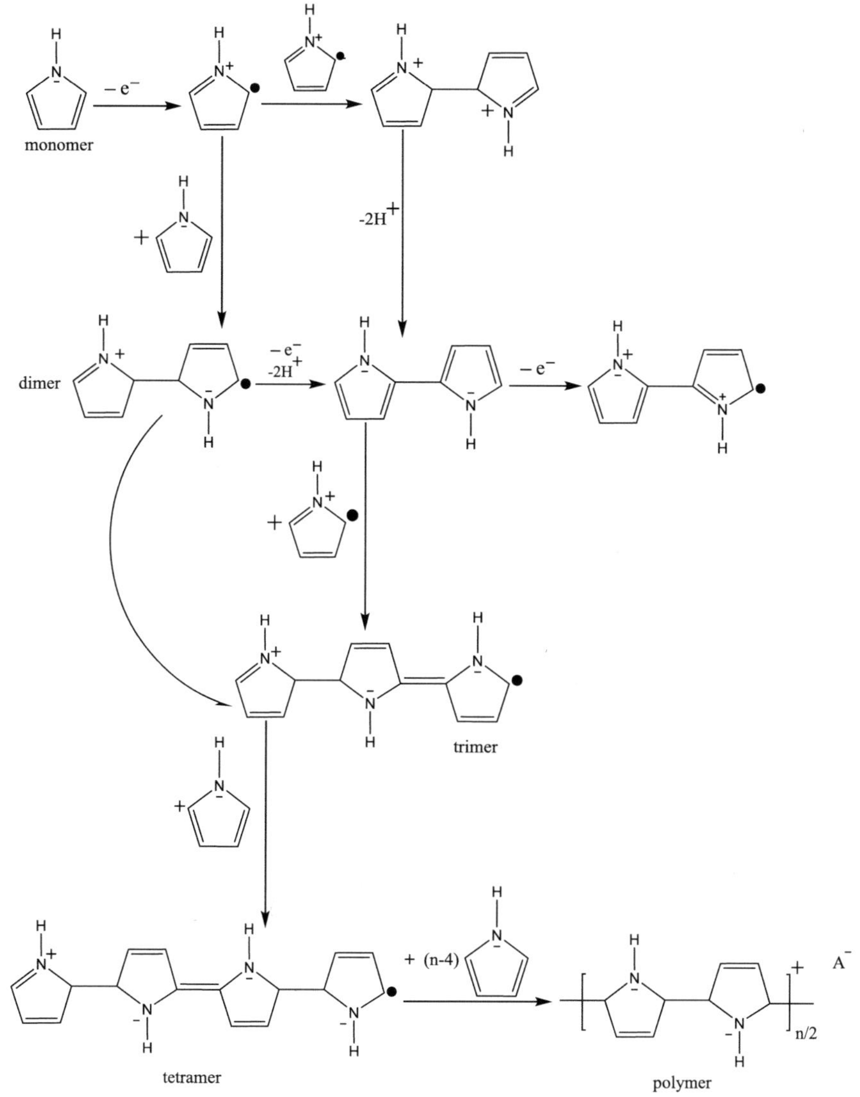

Fig. 1.5 Polypyrrole polymerisation

ulation of surrounding tissue while having a lower mechanical mismatch with tissue when considered as the building blocks of 3D structures. Therefore, there has been a growing interest over the past decade to employ varieties of processing methods for fabrication of multifunctional electroactive core-sheath structures in terms of both fundamental and applied science. Merging modern spinning methods and the novel era of processable organic conductors technologies provide extra functions embedded in fibre structures in addition to traditional properties of the textile fibres [125].

Multiaxial structured fibres not only offer improved characteristics comparing to those of the single components but also adding the advantage of providing a flexible system which may be optimised for a variety of purposes.

Up till now, a huge amount of intellectual effort has been put into the fabrication of mono or bicomponent fibres for ranges of applications through several fabrication methods. Printing and fibre spinning technologies can be regarded as two of the most prominent techniques which are being investigated for the development of devices based on ranges of materials including ICPs, natural polymers and graphene. Printing method has been in the centre of attention in recent years as a fast, old and inexpensive way of fabrication of a number of materials [126]. However, printing approaches are not included in this work. Spinning of polymer fibres has also witnessed great progress over the past few decades as an interdisciplinary field that applies the principles of engineering and material science toward the development of textile substitutes [188]. Spinning is a specialised form of extrusion that uses a spinneret to form multiple continuous filaments or mono filaments. All fibre forming processes-regardless of the materials involved-are irreversible processes involving the rapid and continuous solidification of a liquid with a very restricted size in two directions. The solidification is brought about by the removal of heat and/or solvent by contacting the liquid with a suitable moving fluid, which can be a gas or a liquid. The first step to produce fibres, continuous threadlike filaments with large length-to diameter-ratios or L/Ds typically >5, is to convert the polymer into a processable and spinnable state. Thermoplastic polymers can be converted into the melt-state and melt-spun. Other polymers may be dissolved in a solvent or chemically treated to form soluble or thermoplastic derivatives and subsequently spun *via* wet-spinning, dry spinning or electrospinning. Main traditional spinning approaches used for fabrication of biomedical devices have been introduced briefly in the subsequent sections followed by providing a more detailed description on the novel generation of spinning methods for fabrication of coaxial fibres for bioapplications *via* different methods together with highlighting the advantages and disadvantages of each method.

1.1.2.1 Electrospinning

Electrospinning is a versatile method for the preparation of long, continuous and fine (nano to sub-micron size range) [189, 190] nonwoven polymer mats or fibres known since early the 1930s [191]. Doshi and Renekar were the first researchers in 1995 to report the electrospinning method for producing nanofibres [192]. Electrospinning shares characteristics of both electro-spraying and conventional solution dry-spinning methods [193]. Electrospun fibres possess properties not found in conventional fibres, including high surface to volume ratio, high aspect ratio, controlled pore size and superior mechanical properties [194]. A typical electrospinning setup (Fig. 1.6) consists of a capillary tube or syringe loaded with a polymer solution, a metal collecting screen, and a high voltage supply [191, 195]. The pendant polymeric droplet at the tip of the needle, when subjected to an electric field in the kV range, will deform into a Taylor cone shape and form a liquid jet. This jet under-

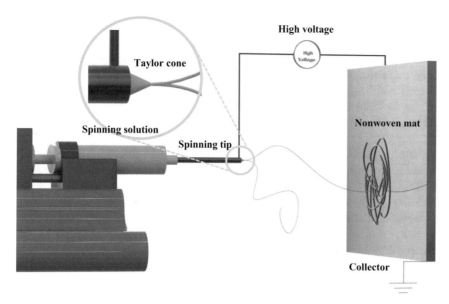

Fig. 1.6 Schematic of electrospinning [195]; Reproduced by permission of The Royal Society of Chemistry (RSC) on behalf of the Centre National de la Recherche Scientifique (CNRS) and the RSC

goes an electrically induced bending instability which results in strong looping and stretching of the jet. Following solvent evaporation, ultrathin fibres are deposited on the collecting screen. Collection systems currently used include a single ground, rotating single ground, dual bar, dual ring, single horizontal ring, etc. which can significantly influence the fibre orientation [196]. Electrospun conductive fibres have found various applications as light emitting diodes (LEDs), chemical and biological sensors, batteries, electromagnetic shielding and wearable electronic textiles (E-textiles) [197].

The morphology and diameters of electrospun fibres are quite versatile depending on a number of parameters including processing including the applied voltage, the distance between spinneret and collector and the feeding rate of polymer solution as well as solution parameters such as intrinsic properties of the spinning solutions, natures of polymer and solvent, polymer molecular weights, conductivity and surface tensions, etc. In addition, different geometries such as highly aligned, yarns and arrays have been achieved using different electrospinning setups. Oriented polyamide nanofibres formation was reported by Dersch et al. as a result of the polymeric jet moving back and forth on the collector [198].

The electrospinning technique has shown to provide non-wovens to the order of few nanometers with large surface areas, ease of functionalisation for various purposes and superior mechanical properties [189]. Also, the possibility of large scale productions as well as simplicity of the process makes this technique very attractive for many different applications. Biomedical field is one of the most widely used application areas among others utilising the technique of electrospinning mainly for development of filtration and protective material, tissue engineering, drug release, wound dressing, enzyme sensors, nanofibre reinforced composites, etc. The versatility in material selection for electrospinning combined with the different ways of method modification, diverse morphological structures with tuned properties for certain applications could be developed. Electrospinning generates loosely connected 3D porous mats with high porosity and high surface area which can mimic ECM structure and therefore makes itself an excellent candidate for use in tissue engineering. Biocompatibility and biodegradability are the other most critical requirements for electrospun fibres as a scaffold which determines the scaffold's ability to degrade within a timeframe in vivo. Drug release and tissue engineering are closely related areas have been targeted for utilization of electrospun fibres. Many research groups have evaluated the properties of the fabricated nanofibres as potential cell supportive tissue scaffolds [189, 198–204]. For instance, PLGA electrospun fibres were fabricated by Ko et al. to be employed as bone-marrow-derived mesenchymal stem cells template which indicated cell attachment and proliferation on PLGA mat [205]. Hsiao et al. also investigated the interaction of primary cardiomyosytes with aligned electrospun PLGA fibres which could direct cells for cardiac tissue [206]. In another study, the conducting polymer of PAni were blended with PCL and gelatin to be electrospun for applying electrical stimulation to nerve stem cells [207]. Electrospinning also affords great flexibility in the development of diverse materials for drug delivery and cell delivery applications [19, 208, 209]. Recent work has examined the possibility of using electrospun matrices as constructs for giving controlled release of a number of drugs including antibiotics [210–212] and anticancer drugs [213, 214] as well as proteins [210, 215, 216] and DNA [217]. Some of the most important biomedical applications of electrospun mats like tissue engineering drug release, wound dressing, enzyme immobilization etc. are highlighted as examples in Table 1.1.

1.1.2.2   Melt-Spinning

Most commercial synthetic fibres are produced by the melt-spinning process. Melt-spinning is a process in which dried polymer granules or chips are melted inside the extruder which is used afterwards as the spinning dope. The obtained filament is quenched and solidified by cooling in a fast fibre solidifying process which is mainly due to the one-way heat transfer [233]. Melt-spinning is considered to be one of the simplest methods compared to other fibre manufacturing methods due to the absence of problems associated with the use of solvents [234]. It is, therefore, the preferred method for spinning many polymers provided the polymer gives a stable melt [235]. A schematic of melt-spinning process is presented in Fig. 1.7.

Melt-spinning has not been considered as a method of choice for development of biostructures, though due to the limitations involved with this method. PLA is one of the most widely spun polymers *via* melt-spinning for bioapplications. However, it was shown that main challenges of PLA foaming are low melt strength and slow crystallization kinetics. The low cell strength leads to cell coalescence and cell rup-

Table 1.1 Some of the representative electrospun systems studied for biomedical applications

Electrospun Mat	Bioapplication	References
PCL	Controlled delivery of Dipyridamole	[218]
Gelatin	Cell regeneration and proliferation for tissue engineering	[219]
PLGA/Silver	antimicrobial wound dressing	[220]
Chitosan/PEO/BG	Bone tissue engineering	[221]
PVA	Drug delivery of Tenofovir for vaginal applications	[222]
Alg/PEO	Wound dressing and sutures	[223]
PU	Wound dressing	[224]
PLLA/HA/Col	Stem cell based therapies in bone tissue engineering	[225]
PCL/Cellulose	Liquid biofilters and biosensor	[226]
PAN	Water disinfection filters	[227]
PCL:PBS	Ofloxacin (OFL) loaded drug delivery system for treatment of ocular infections	[228]
PCL/Col	Smooth muscle cells growth for blood vessel engineering	[229]
PCL	Contractile cardiac grafting	[230]
PLGA	Tissue engineering which supports and guide cell growth	[231]
Gelatin/Silver	Antibacterial activity wound dressing	[232]

ture during growth. There also exist few reports of the melt-spinning of conducting polymer fibres due to some major limitations. These include decomposition at temperatures below the melting point, poor control over the exact temperature of the polymer melt during spinning, thermo-mechanical history of the melt, and final fibre structure [236].

1.1.2.3 Wet-Spinning

Of all the fibre spinning methods, solution spinning methods have the longest history. Wet-spinning was one of the original methods for producing synthetic fibres and was first used in the late 19th century [237]. In wet-spinning, the polymer dissolved in a suitable solvent is extruded directly into a coagulation bath containing a liquid which is miscible with the spinning solvent but a non-solvent of the polymer. This leads to solvent removal from the polymer and solidification of the fibre as precipitation

Fig. 1.7 Schematic of
melt-spinning [195];
Reproduced by permission
of The Royal Society of
Chemistry (RSC) on behalf
of the Centre National de la
Recherche Scientifique
(CNRS) and the RSC

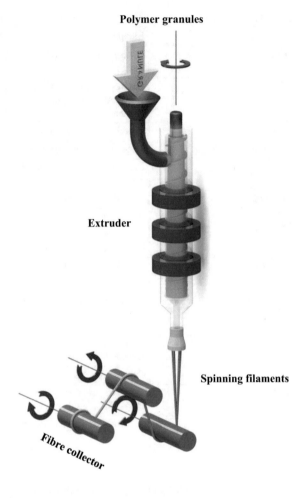

occurs. Wet-spinning involves mass transfer of the solvent and non-solvent for fibre solidification, which is slower compared to the heat transfer process of cooling associated with melt-spinning, and to the evaporation associated with dry spinning [234]. It is usually subdivided into three main steps based on different spinning strategies as follows; (a) phase separation, (b) gel separation and (c) liquid crystal spinning [238]. (a) During the phase separation, rapid formation of the fibre structure will occur as a result of polymer solution exposure with the coagulation bath. As the polymer fluid is injected into the non-solvent, the solvent is extracted from the polymer solution causes the polymer to be precipitated in the bath to form a semi-solid fibre. Further solidification into a coagulation bath provides sufficient cohesion and strength for the fibre to be continuously collected when coming out of the coagulation bath. In the second step, the polymer is coagulated due to intermolecular bonds such as ionic cross-linking by a salt or another reacting agent. In the liquid crystal spinning stage,

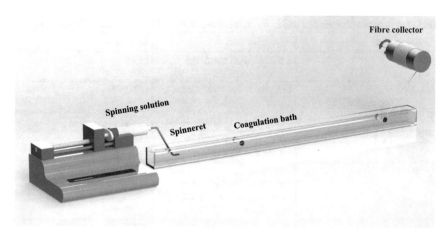

Fig. 1.8 Schematic of a Lab scale wet-spinning line [195]; Reproduced by permission of The Royal Society of Chemistry (RSC) on behalf of the Centre National de la Recherche Scientifique (CNRS) and the RSC

lyotropic crystalline solution provides sufficient alignment and cohesiveness to form a solid crystalline phase for fibres. A schematic of wet-spinning is shown in Fig. 1.8.

Like electrospinning, wet-spun microfibres have gained considerable interest in a number of biomedical applications such as scaffolding substrates as well as drug delivery systems over the past decade [239–242]. Flexibility, appropriate mechanical properties, versatility in choice of materials and structures and their the ability to be fabricated into devices using weaving technique such as knitting/braiding methods provide promising potentials for wet-spun fibres to be utilised for biomedical purposes. The ability to precisely control the attachment, migration, alignment, proliferation and differentiation of cells is also extremely important for the regeneration of tissues that require a complex sequence of biological cues and structural support. To date, wet-spun microfibres have been investigated for a range of polymers as tracks to guide and direct the behaviour of cells for a variety of applications, such as vascular tissue engineering [243], musculoskeletal tissue engineering [244] and wound healing [245]. The majority of studies involving wet-spun microfibre scaffolds have primarily focused on the release of therapeutics or biocompatibility and tissue regeneration capabilities of 3D scaffolds. Lee et al. fabricated sub-micron fibres of PLGA [246] as platforms for tissue regeneration *via* precipitating a solution of PLGA/DMSO into a bath of glycerol-containing water and studied the cell attachment and growth on them to be applied as a nerve guide. PLGA wet-spun fibres were also formed into a tissue scaffold have been also employed as appropriate substrates for ligament replacement due to the characteristics it offers such as mainly high biocompatibility and mechanical properties [247]. Razal et al. wet-spun hyaluronic acid and Chit with CNT to form a composite fibre which has indicated to support the growth of L-929 fibroblast cells [248].

Table 1.2 Some of the representative wet-spun systems studied for biomedical applications

Wet-spun fibre	Bioapplication	References
Silk-PU	Small-diameter artificial vascular scaffolds	[240]
HA–CNT	Directed nerve and/or muscle repair	[248]
Chit-PCL	Scaffolds for cartilage repair	[255]
Col	Scaffolding regeneration of tendon and ligament tissue	[256]
Corn starch-PCL	Designing osteoconduc-tive/osteoinductive 3D structures for bone tissue engineering	[257]
PAN	Drug delivery system for controlled release of Tamoxifen citrate	[258]
Amyloid	Bone tissue engineering	[259]
Alg-Silk	Wound dressings	[260]
Chit	Biomedical scaffolds	[261]
Starch	Tissue engineered scaffolds	[242]
PAN-PCL	Ibuprofen drug sustained release	[262]
Alg-Chit	Scaffolds for ligament and tendon tissue engineering	[263]

From a drug-delivery perspective, wet-spinning is most similar to conventional microsphere-based drug encapsulation techniques and avoids the potential for thermal denaturation of therapeutics, unlike melt-spinning and dry-spinning methods. Thus, it is not surprising that a broad range of therapeutics such as antibiotics [122], glycosaminoglycans [249], proteins [250], growth factors [251], genes [252] and viruses [253], have been successfully incorporated into microfibres produced by wet-spinning methods and other variants. Dry-wet-spun microfibres have even been approved as ocular drug-delivery devices for treatment of posterior-segment diseases, such as macular oedema [254]. Some of the most important biomedical applications of electrospun mats like tissue engineering drug release, wound dressing, enzyme immobilization etc. are highlighted as examples in Table 1.2.

1.1.2.4 Dry-Spinning

Dry-spinning is another type of solution spinning which was first employed around the same time as wet spinning [264]. This old method for the preparation of synthetic fibres has many basic principles in common with wet spinning, including the requirement that the polymer needs to be dissolved in a solvent. Compared to wet spinning, solidification is achieved more easily through evaporation of the solvent, which must be highly volatile, and without requiring a coagulation bath. Dry-spinning is suitable for polymers which are vulnerable to thermal degradation, cannot form viscous melts, and when specific surface characteristics of fibres are required [266]. It is the preferred method for polyurethane, polyacrylonitrile, and fibres based on ophthalamide, polybenzimidazoles, polyamidoimides, and polyimides due to bet-ter physicomechanical fibre properties [265]. Dry-spinning of continuous cellulose fibres from a bio-residue of bleached banana rachis waste has been lately reported [266]. Recently, Zhang et al. also fabricated a hybrid dry-spun fibres from regenerated silk fibroin/graphene oxide aqueous solution using a dry-spinning approach [267].

1.1.3 Manufacturing Processes for Fabrication of Coaxial Biofibres

Advanced fibre processing techniques such as coaxial or triaxial spinning methods have had recently attracted a considerable deal of attention to producing coaxial fibres from natural and synthetic polymers for a range of biomedical purposes. Coaxial electrospinning is the simplest approach in the field of coaxial spinning which has been extensively exploited to generate hollow and core-sheath nanofibres [268–272]. However, electrospinning can produce extremely fine fibres in the form of a non-woven mat; however, the mechanical testing of individual fibres is not feasible. There have been also reported on difficulties involved with the preparation of fibres using already charged polymer backbones *via* electrospinning wherein a stable jet cannot be achieved and no nanofibres would form as a result, although single droplets may be achieved (electrospray). In addition to this, there is a limitation in choosing the maximum concentration for a given solution by which it could flow. Consequently, the molecular weight and the concentration as long as a solvent with the necessary volatility which can be spun this way, are within a certain range. Thus, although this method shows a lot of promise, these restrictions are placed on the spinnability of certain polymers by solution parameters like viscosity and surface charging.

Recently, wet-spinning has been utilised to produce fibres with similar structures to coaxial fibres [1, 122, 273, 274]. Despite those preliminary studies there are only a few reports in the literature of fabricating coaxial fibres using coaxial spinnerets using wet-spinning. Recently, Liang Kou et al. have also reported on the production of carboxyl methyl cellulose (CMC)-wrapped graphene/CNT coaxial fibres for super-capacitor applications [275]. Therefore, it is still a big challenge to develop a simple yet effective wet-spinning approach to directly prepare sheath-protected fibres as electrodes. The key principle in wet-spinning is to make polymer fibres by transition from a soluble to a non-soluble phase as discussed in Sect. 1.1.2.3 [276]. The major difference between conventional wet-spinning with the coaxial method is that in the coaxial process, two different polymer solutions are injected into a coaxial spinneret together and are co-extruded into a bath while retaining a coaxial structure. Using this method, long uniformly shaped fibres can be produced in a process which allows great flexibility in the composition of the fibre components. However, determining features are placed within a wide range from the material requirements such as optimal viscosities and concentrations of spinning solutions, solvent/solution miscibility, material surface charges, etc. as well as interaction between materials/coagulation bath (numbers of chemical/physical reactions between spinning components and the bath), time restrictions for storage of fibres in the bath and several controlling process parameters such as core/sheath injection rates, drawing ratio, spinneret design, post-treatment procedures and so on. It is worth noting that as the sheath and the core solutions are in contact and undergo the same procedure, the degree of dissimilarity between them, in terms of composition, and physical and rheological properties is critical in the formation of the coaxial structure.

1.1.3.1   Methods Originated from Electrospinning

In recent years, many modifications have been made in the basic electrospinning process in order to enhance the quality and improve the functionality of the resulting nanofibre structures. One such modification that has gained much attention and holds great promise in a variety of applications is the preparation of coaxial bicomponent nanofibre structures using "co-axial electrospinning" which is also called "two-fluid electrospinning" [196]. In this process, two dissimilar materials are delivered independently through a coaxial spinneret and drawn using high voltages to generate nanofibres in core-sheath configuration [277]. Many different polymers have been electrospun relatively successfully in a solution or melt form. These have included synthetic polymers (polyesters, polyamides, polyurethanes, polycarbonates, polysulfones, etc.), natural materials (collagen, gelatin, elastin, chitosan, silk etc.) and synthetic biodegradable polymers (polyglycolic acid, polycaprolactone, polylactic acid, polylactide-co-glycolide, etc.). In the present time, there are three common processes to produce coaxial fibres *via* methods with similar principals to that of electrospinning including coaxial electrospinning, coaxial melt-electrospinning and emulsion electrospinning.

1.1.3.1.1   Coaxial Electrospinning

This method was established in 2006 which bifurcated from previously described electrospinning in Sect. 1.1.2.1. The general set up is quite similar to that used for electrospinning which employs electric forces acting on polymer solutions in dc electric fields and resulting in significant stretching of polymer jets due to a direct pulling and growth of the electrically driven bending perturbations [199, 278]. A modification is made in the spinneret by inserting a smaller (inner) capillary that fits concentrically inside the bigger (outer) capillary to make coaxial configuration as shown in Fig. 1.9.

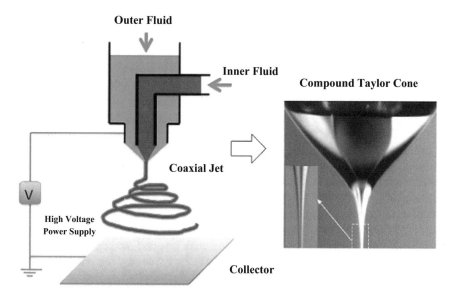

Fig. 1.9 A schematic for coaxial electrospinning set up [279]

In coaxial electrospinning, the outer needle is attached to the reservoir containing the sheath solution and the inner is connected to the one holding the core solution. Syringes with two compartments containing different polymer solutions would undergo high voltages leading them to be charged. The charge accumulation occurs predominantly on the surface of the sheath liquid coming out of the outer co-axial capillary [200]. Once the charge accumulation reaches a certain threshold value due to the increased applied potential, a fine jet extends from the cone due to the electrostatic repulsion of the charges in the polymer liquids. The coaxial set up expectedly requires a carefully designed and manufactured co-axial spinneret. Due to the similarities between coaxial electrospinning process to that of the conventional electrospinning, all variables that govern the quality of the process and the morphology of the fibres in the latter also affects the behaviour in the former. Those affecting factors are quite diverse comprising both material (viscosities, concentrations, solvent/solution miscibility and incompatibility, solvent vapour pressure, solution conductivities, etc.) and process parameters (applied voltage, flow rates, spinneret design, etc.). Additionally, as the sheath and the core solutions are in contact and undergo the same bending instability and whipping motion, the degree of dissimilarity between them, in terms of composition, and physical and rheological properties, plays an important role in the formation of the composite fibre.

Coaxial electrospinning rapidly gained popularity and was implemented by a number of groups. The method has provided many benefits over the other conventional spinning such as degradation of an active encapsulant can be significantly minimized or avoided [182, 280–282]. Also, it has been noted that burst release of a core is effectively minimized in case studies of release kinetics. Burst release is a phenomenon commonly observed in delivery devices of different forms and compositions. The burst effect may be favourable for certain *drug administration strategies* or applications such as wound treatment, encapsulated flavours, targeted delivery and pulsatile release [283]. However, burst release is also likely to cause negative effects such as local/systemic toxicity, short in vivo half-life, economically inefficiency and a shortened release profile. Burst release is often associated with device geometry, surface characteristics of the host material, heterogeneous distribution of drugs within the polymer matrix, intrinsic dissolution rate of the drug, heterogeneity of matrices (pore density), etc. The main objective of drug delivery systems is to achieve an effective therapeutic administration *via* a sustained drug release over an extended period of time. Coaxial electrospun fibres have exhibited more steady release rate behaviour due to the creation of specific geometries. In addition to the release profile of biomolecules, mechanical properties and biocompatibility could also be enhanced *via* coaxial electrospinning to provide additional functional properties [279, 284, 285]. Zhang et al. reported the coaxial electrospinning of PCL-r-Gelatin bi-component nanofibres for tissue engineering applications [286]. Recently, James et al. described a coaxial electrospinning method for development of novel core–shell nanofibres using gelatin as the core material and chitosan as shell as tissue scaffold matrices [287].

1.1.3.1.2 Emulsion Electrospinning

Emulsion electrospinning is similar to the normal solution electrospinning, except that the solution is replaced by an emulsion. Jets are generated from the emulsion liquid and stretched into ultrafine fibres. The dispersed drop in the emulsion turns into the core of the electrospun fibres, and the continuous matrix becomes the shell. The water-in-oil emulsion electrospinning is particularly used for encapsulating hydrophilic drugs or bioactive molecules inside the core of electrospun core/shell fibres to avoid burst release and prolong the release time [208, 209, 288, 289].

1.1.3.1.3 Coaxial Melt-Electrospinning

Coaxial melt-electrospinning is quite similar to the traditional electrospinning method with a small difference of using polymer melts instead of solutions, for which a heating system is used that surrounds the reservoir [196]. The polymer melt is usually produced by heating from either resistance heating, circulating fluids, air heating or lasers. This so-called coaxial melt-electrospinning is the result of the combination of melt-electrospinning with coaxial spinneret which provides a facile method for the encapsulation of solids in a composite or polymer matrix and extends the technique to create new morphologies and architectures. Melt electrospinning was first reported in the 1980s but has not been studied as extensively as solution electrospinning due to the more expensive setup needed to maintain elevated temperature of the melt and the limitation of the low conductivity and high viscosity

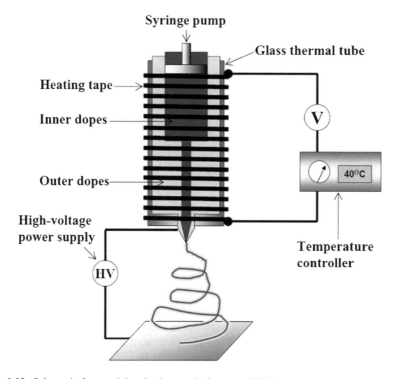

Fig. 1.10 Schematic for coaxial melt-electrospinning setup [291]

of polymer melts in conventional melt electrospinning setups to achieve significant fibre diameter reduction by electrostatic forces [290]. A schematic of coaxial melt-electrospinning setup is shown in Fig. 1.10.

Melt electrospinning has several distinct advantages over its solution-based counterpart, most notably environmental friendliness and low cost due to the absence of a solvent [292]. There is also the particular advantage when using materials with different melting points such as coaxial nanofibres need to be incorporated for phase change applications. Processing of fibres from melts is a widely used method in commercial fibre fabrication. Using the melt electrospinning technique, commodity polymers such as poly propylene (PP) and poly ethylene terephthalate (PET) have been directly drawn into nonwoven fibres [292]. However, coaxial melt electrospinning has not been reported to be used for fabrication of biofibres to the best knowledge of the author.

1.1.3.2 Methods Descended from Wet-Spinning

Although coaxial electrospinning and its earlier discussed derived approaches were applied by many researchers to produce coaxial fibres, only a few reports appeared in the literature reported the successful fabrication of hybrid fibres *via* coaxial wet-spinning methodology to the knowledge of the author. This might be due to the complexity of this method because of the plurality of parameters involved in the successful formation of a core-sheath structure inside a coagulation bath. In the first instance, the fabrication of wet-spun coaxial fibres appears a straightforward task; two different components are injected through a coaxial spinneret at once into a proper coagulation bath to form a coaxial structure. However, this simple approach presents several challenges. Many parameters are needed to be controlled and regulated in order to hold both components in a coaxial structure. Among those, the solution properties are known as the key factors affecting the spinning process including mainly the material concentrations, viscosities, surface charges, surface tensions, polymer natures and functionalities and so on. However, less important considerations such as systematic variables (injection rates (V_i), core to sheath injection rates ratio, take-up velocity (V_t), coagulation bath constituents, drawing velocity (V_d), spinneret specialties, post-treatment processes, etc.) and ambient conditions (temperature, post-spinning conditions) could be also influenced the process as well as final fibre properties significantly. Up to date, different examples of coaxial fibres have been reported using a number of materials such as conducting polymers, metals, natural polymers and carbon-based components *via* various production methods. Most efforts have been focussed on approaches based on electrospinning to date to produce coaxial structures. Some of the main techniques to produce coaxial fibres and yarns are described in the following sections. However, among earlier efforts to produce wet-spun coaxial fibres, few methods could be found sharing similar procedures to that of coaxial wet-spinning described below.

1.1.3.2.1 Production of Hollow Fibres

Many different types of semipermeable hollow fibres have been prepared by means of the melt, dry or wet-spinning techniques [1, 293, 294]. The procedure of achieving a hollow structure is quite similar to that of coaxial wet-spinning in which the sheath spinning solution was delivered from a chamber to the external spinneret nozzle through injection, whereas the core fluid is usually substituted simply with pressurized water or the coagulant fluid injected to the central nozzle. These fibres are being applied for various purposes such as gas separation, ultrafiltration, reverse osmosis as well as many biological applications including drug delivery, dialysis and tissue engineering.

First attempts to produce hollow fibres *via* wet-spinning might go back to the year 1976 when Cabasso et al. presented a method to develop porous polysulfone hollow fibres as supports for ultrathin membrane coatings useful in high-pressure reverse osmosis processes [293]. Later on, Aptel and his co-workers reported of producing asymmetric hollow fibres from a three-component dope mixture containing polysulfone, polyvinylpyrrolidone (PVP), and N,N-dimethylacetamide using wet-dry-spinning technique [295]. Then, another research group produced polymer hollow fibre membranes for removal of toxic substances from blood in 1989 [296].

Delivery systems with diverse release profiles spanning from a few days to several months have been achieved by encapsulation of biological molecules as well as drug reservoirs into the core of hollow fibres or chemically crosslinking or adsorbing therapeutics to the surfaces of fibres. With regard to the encapsulation of drugs within wet-spun filaments, a critical issue is obtaining the appropriate release characteristics and mechanical integrity for specific cell type/tissue architecture. Phillips et al. have reported of fabrication of PVDF hollow fibre for drug delivery to tumour cells. Polacco and his colleagues also described a hollow PLGA fibre fabrication as a controlled drug release system [297]. Lee et al. have described a new method for encapsulation of human hepatocellular carcinoma (HepG$_2$) cells in wet-spun chitosan-alginate microfibres [298]. In addition to those mentioned, many other research groups reported on using a hollow fibre structure for delivering biomolecules [299–303].

1.1.3.2.2 Wet-Spun Core-Skin Fibres *via* Coating

Preparation of so-called core-skin fibres is another method ending up to similar structures to that of coaxial wet-spun fibres which was applied by several research groups, so far. Actually, this approach is same as the traditional wet-spinning method. However, the coagulation bath which the polymer participates in is altered with a second material. Consequently, a thin layer of the second component is formed on fibres like a skin. A few fibre companies have recently reported on the production of skin-core fibres. Mitsubishi Rayon Co. Ltd. has succeeded in the development of series of core-sheath multifunctional acrylic fibres whose core is filled with different materials using a wet process in 2006 [304]. A. Niekraszewicz et al. also reported of producing alginate-chitosan fibres characterised by very high water retention values of up to 1300% makes them suitable for use in sanitary and medical products [305]. Later on, A. Granero et al. described a method for fabrication of CNT—biopolymer core-skin fibres thru injecting either a dispersion of CNT into a coagulation bath based on carrageenan and chitosan with opposite charge [273]. Recently, another researcher has taken a step forward by loading with an antibiotic drug of ciprofloxacin hydrochloride into PEDOT: PSS as the inner core to the electropolymerised outer shell layer of Ppy [122].

1.1.3.2.3 Coaxial Wet-Spinning

Despite those preliminary studies to produce fibres with similar structures to that of the coaxial fibres such as hollow and core-skin fibres mentioned previously [1, 122, 273, 274], there are only a few reports in the literature of fabrication of coaxial fibres using a coaxial spinneret for wet-spinning [275]. The major difference of conventional wet-spinning with the coaxial method is that in the coaxial process, two different polymer solutions are injected into a coaxial spinneret together and are

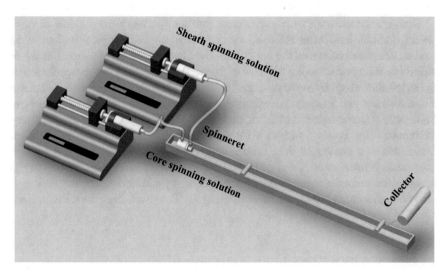

Fig. 1.11 Coaxial wet-spinning setup

co-extruded into a bath while retaining a coaxial structure. A schematic of coaxial wet-spinning setup is shown in Fig. 1.11.

Coaxial wet-spinning produces hollow or core-shell fibres that can be used for quite a lot of purposes such as controlled release applications, electronic textiles, sensors and actuators [306, 307]. As a matter of fact, for the successful production of coaxial fibres, several parameters are needed to be controlled and regulated. Thus, development of a simple yet effective wet-spinning approach for directly preparation of sheath-protected fibre electrodes has still remained a challenge. Among those, solution properties are of great importance such as material concentrations, viscosities, surface charges, surface tensions, polymer natures and functionalities and etc. However, process parameters could also influence the process as well as the final fibre properties significantly. G. Park and his co-workers were successful to spin CNT/Poly (vinyl alcohol) fibres with a sheath-core structure *via* wet-spinning [308]. Recently, Liang Kou et al. have also reported on the production of CMC/wrapped graphene/CNT coaxial fibres for supercapacitor applications [275].

1.2 A Brief Overview of Fibre-Based Scaffolds

Applications of CPFs in biological field were expanded later on with the discovery that these materials were compatible with many biological molecules in the late 1980s [113]. Most CPs present a number of important advantages for biomedical applications, including biocompatibility, ability to entrap and controllably release biological molecules, ability to transfer charge from a biochemical reaction, and the potential to alter the properties of the CPs to better suit the nature of the specific application [113]. Conducting fibres can provide self-supporting three-dimensional, flexible structures suitable for in vitro and in vivo bionic applications compared to the films. These functional aspects may also require the overlap of certain characteristics for example for uses in implantable batteries and bio-actuators [113, 309]. In more detail, storage or conversion of energy and provide the required biocompatibility. Today, the major bioapplications of CPFs are generally within the area of electrical stimulation and signal recording [117, 310, 311], drug-delivery devices [122], tissue-engineering scaffolds [114, 154, 312], and biosensors [112, 113]. Recently, there is a growing interest in using conducting fibres for neural tissue engineering applications. These conductive fibrillar pathways may provide appropriate replacements for nerve fibres after injuries. Electrical stimulation has been shown to enhance the nerve regeneration process and this consequently makes the use of electrically conductive polymer fibres very attractive for the construction of scaffolds for nerve tissue engineering. For instance, Li et al. investigated the feasibility to generate novel electrospun PAni-gelatin blended scaffolds as potential scaffolds for neural tissue engineering [313]. They reported that as-prepared fibers are biocompatible, supporting attachment, migration, and proliferation of rat cardiac myoblasts. In another study, the feasibility of fabricating a blended fibre of PAni–polypropylene as a conductive pathway was studied for neurobiological applications [312]. In addition, production of conducting fibres for controlled drug release applications is currently of particular interest of many research groups. Fabrication of PEDOT:PSS-chitosan hybrid fibres was described using a novel wet spinning strategy to achieve a controlled release of an antibiotic drug [122]. Still, there remain limitations for use of CPs due to their manufacturing costs, material inconsistencies, poor solubility in solvents and inability to directly melt process. Moreover, oxidative dopants could diminish their solubility in organic solvents and water and hence their processability.

1.3 Thesis Objectives

This thesis aims to develop coaxial and triaxial electroactive fibres to be potentially used as electrodes for biomedical applications. In this study, a novel wet spinning approach is developed that employs a coaxial spinneret for production of core-sheath fibres which could potentially provide a mechanism for the safe electrical stimulation of tissue whilst also avoiding undesirable cell damage. The electroactive nature of organic conductors such as graphene and conducting polymers enables improved mechanical and electrical properties for electrical stimulation purposes, while the biofriendly character of hydrogels would be beneficial in enhancing the electro-cellular communication process.

Among the many production techniques available, advanced fibre processing methods such as coaxial and triaxial spinning have attracted a great deal of attention. Coaxial electrospinning which has been extensively utilised to generate hollow and core-sheath nanofibres, showed some deficiencies such as unfeasibility of testing the mechanical properties of individual fibres, difficulties for fibres preparation from charged polymer backbones and limitations for using viscous spinning solutions as well as selection of the solvent. Thus, although this method shows a lot of promise, these restrictions are placed on the spinnability of certain polymers by solution parameters like viscosity and surface charging. Recently, wet-spinning has been utilised to enable a simple yet effective approach to directly prepare sheath-protected fibres. The fabrication of wet-spun coaxial fibres appears a straightforward task at first glance; however, many parameters are needed to be controlled and regulated in order to hold both components in a coaxial arrangement which mainly include the solution parameters combined with process variables. Coaxial electrospinning was applied by many researchers to produce coaxial fibres; however, only a few reports described the successful fabrication of hybrid fibres *via* coaxial wet-spinning methodology to the knowledge of the author. This might be due to the complexity of this method because of the plurality of parameters involved in the successful formation of a core-sheath structure inside a coagulation bath.

Specifically, we have developed a one-step wet-spinning method to produce coaxial fibres using a couple of conductors and hydrogels via a facile continuous technique. Subsequently, we tried CVD technique which facilitates deposition of another organic conductor on the surface of coaxial fibres for the structural simulation of battery structures. Spinnability of coaxial fibres considering spinning conditions such as appropriate choice of hydrogel, appropriate concentrations and viscosity, injection flow rate of each spinning solution, identification of suitable reducing agent selection and reduction method have been investigated in this thesis. As-prepared fibres were also characterised in terms of mechanical, electrical, electrochemical, swelling and biological properties to be optimized for use in biomedical applications.

References

1. Chwojnowski A, Wojciechowski C (2009) Polysulphone and polyethersulphone hollow fiber membranes with developed inner surface as material for bio-medical applications. Biocybern Biomed Eng 29:47–59
2. Dash M, Chiellini F, Ottenbrite RM, Chiellini E (2011) Chitosan—a versatile semi-synthetic polymer in biomedical applications. Prog Polym Sci 36:981–1014. https://doi.org/10.1016/j. progpolymsci.2011.02.001
3. Chwojnowski A, Wojciechowski C, Dudzin'ski K, Łukowska E (2009) Polysulphone and polyethersulphone hollow fiber membranes with developed inner surface as material for biomedical applications. Biocybern Biomed Eng 29:47–59
4. Descoteaux C, Provencher-Mandeville J, Mathieu I, Perron V, Mandal SK, Asselin E, Berube G (2003) Synthesis of 17β2-estradiol platinum(II) complexes: biological evaluation on breast cancer cell lines. Bioorg Med Chem Lett 13:3927–3931. https://doi.org/10.1016/j.bmcl.200 3.09.011
5. Ducheyne P, Qiu Q (1999) Bioactive ceramics: the effect of surface reactivity on bone formation and bone cell function. Biomaterials 20:2287–2303. https://doi.org/10.1016/S0142-9 612(99)00181-7
6. Ramaswamy Y, Wu C, Zhou H, Zreiqat H (2008) Biological response of human bone cells to zinc-modified Ca-Si-based ceramics. Acta Biomater 4:1487–1497. https://doi.org/10.1016/j. actbio.2008.04.014

7. Rae T (1986) The biological response to titanium and titanium-aluminium-vanadium alloy particles. II. Long-term animal studies. Biomaterials 7:37–40. https://doi.org/10.1016/0142-9612(86)90086-4

8. Wever DJ, Veldhuizen AG, Sanders MM, Schakenraad JM, Van Horn JR (1997) Cytotoxic, allergic and genotoxic activity of a nickel-titanium alloy. Biomaterials 18:1115–1120. https://doi.org/10.1016/S0142-9612(97)00041-0

9. Kimura Y, Hokugo A, Takamoto T, Tabata Y, Kurosawa H (2008) anterior cruciate ligament regeneration by biodegradable scaffold combined with local controlled release of basic fibroblast growth factor and collagen wrapping. Tissue Eng 14:47–57. https://doi.org/10.1089/tec.2007.0286

10. Osada BY, Gong J (1998) Soft and wet materials: polymer gels. Adv Mater 10:827–837

11. Swann JMG, Ryan AJ (2009) Chemical actuation in responsive hydrogels. Polym Int 58:285–289. https://doi.org/10.1002/pi.2536

12. Drury JL, Mooney DJ (2003) Hydrogels for tissue engineering: scaffold design variables and applications. Biomaterials 24:4337–4351. https://doi.org/10.1016/S0142-9612(03)00340-5

13. Ionov L (2014) Hydrogel-based actuators: possibilities and limitations. Biochem Pharmacol 17:494–503. https://doi.org/10.1016/j.mattod.2014.07.002

14. Jager EWH (2012) Actuators, biomedicine, and cell-biology. In: Proceedings of SPIE, pp 1–10

15. De SK, Aluru NR, Johnson B, Crone WC, Beebe DJ, Moore J (2002) Equilibrium swelling and kinetics of pH-responsive hydrogels: models, experiments, and simulations. J Microelectromech Syst 11:544–555

16. Jain JL, Sunjay J, Nitin J (2005) Polysaccharides. In: Fundamentals of biochemistry, pp 114–131

17. Martínez A, Fernández A, Pérez E, Benito M, Teijón JM, Blanco MD (2010) Polysaccharide-based nanoparticles for controlled release formulations. In: The delivery of nanoparticles. InTech, pp 185–222

18. He B, Leung M, Zhang M (2010) Optimizing creation and degradation of chitosan-alginate scaffolds for in vitro cell culture. J Undergrad Res Bioeng 31–35

19. Malafaya PB, Silva GA, Reis RL (2007) Natural-origin polymers as carriers and scaffolds for biomolecules and cell delivery in tissue engineering applications. Adv Drug Del Rev 59:207–233. https://doi.org/10.1016/j.addr.2007.03.012

20. Nair LS, Laurencin CT (2007) Biodegradable polymers as biomaterials. Prog Polym Sci 32:762–798. https://doi.org/10.1016/j.progpolymsci.2007.05.017

21. Wang J, Huang X, Xiao J, Yu W, Wang W, Xie W, Zhang Y, Ma X (2010) Hydro-spinning: a novel technology for making alginate/chitosan fibrous scaffold. J Biomed Mater Res A 93:910–919. https://doi.org/10.1002/jbm.a.32590

22. Dutta P, Rinki K, Dutta J (2011) Chitosan: a promising biomaterial for tissue engineering scaffolds. Chit Biomater II 244:45–80. https://doi.org/10.1007/12_2011_112

23. Suh J, Matthew H (2000) Application of chitosan-based polysaccharide biomaterials in cartilage tissue engineering: a review. Biomaterials 21:2589–2598

24. El-Tahlawy K, Hudson S (2006) Chitosan: aspects of fiber spinnability. J Appl Polym Sci 100:1162–1168. https://doi.org/10.1002/app.23201

25. Stanford ECC (1881) Improvements in the manufacture of useful products from seaweeds. Br Pat 142

26. Bansal V, Sharma P, Sharma N (2011) Applications of chitosan and chitosan derivatives in drug delivery. Advan Biol Res 5:28–37

27. Moshaverinia A, Ansari S, Chen C, Xu X, Akiyama K, Snead ML, Zadeh HH, Shi S (2013) Biomaterials co-encapsulation of anti-BMP2 monoclonal antibody and mesenchymal stem cells in alginate microspheres for bone tissue engineering. Biomaterials 34:6572–6579. https://doi.org/10.1016/j.biomaterials.2013.05.048

28. Quigley AF, Bulluss KJ, Kyratzis ILB, Gilmore K, Mysore T, Schirmer KSU, Kennedy EL, O'Shea M, Truong YB, Edwards SL, Peeters G, Herwig P, Razal JM, Campbell TE, Lowes KN, Higgins MJ, Moulton SE, Murphy MA, Cook MJ, Clark GM, Wallace GG, Kapsa RMI

(2013) Engineering a multimodal nerve conduit for repair of injured peripheral nerve. J Neural Eng 10:1–17. https://doi.org/10.1088/1741-2560/10/1/016008

29. Costa-Pinto A, Reis R, Neves N (2011) Scaffolds based bone tissue engineering: the role of chitosan. Tissue Eng B 17:331–347. https://doi.org/10.1089/ten.teb.2010.0704

30. Qin Y (2008) Review-alginate fibres: an overview of the production processes and applications. Polym Int 57:171–180. https://doi.org/10.1002/pi.2296

31. Mirabedini A, Foroughi J, Romeo T, Wallace GGGG (2015) Development and characterization of novel hybrid hydrogel fibers. Macromol Mater Eng 300:1217–1225. https://doi.org/1 0.1002/mame.201500152

32. Qin Y (2005) Ion-exchange properties of alginate fibers. Text Res J 75:165–168. https://doi. org/10.1177/004051750507500214

33. Lansdown ABG (2006) Silver in health care: antimicrobial effects and safety in use. Curr Probl Dermatol 33:17–34. https://doi.org/10.1159/000093928

34. Qin Y (2005) Silver-containing alginate fibres and dressings. Int wound J 2:172–176. https:// doi.org/10.1111/j.1742-4801.2005.00101.x

35. Walker M, Parsons D (2014) The biological fate of silver ions following the use of silver-containing wound care products—a review. Int wound J 11:496–504. https://doi.org/10.111 1/j.1742-481X.2012.01115.x

36. Qin Y (2004) Gel swelling properties of alginate fibers. J Appl Polym Sci 91:2–6

37. Qin Y (2008) The gel swelling properties of alginate fibers and their applications in wound management. Polym Adv Technol 19:6–14. https://doi.org/10.1002/pat.960

38. Agboh O, Qin Y (1998) Chitin and chitosan fibers. Polym Adv Technol 8:355–365

39. Khor E, Yong L (2003) Implantable applications of chitin and chitosan. Biomaterials 24:2339–2349. https://doi.org/10.1016/S0142-9612(03)00026-7

40. Kumar M (1999) Chitin and chitosan fibres: a review. Bull Mater Sci 22:905–915

41. Sonia T, Sharma C (2011) Chitosan and its derivatives for drug delivery perspective. Adv Polym Sci 243:23–54. https://doi.org/10.1007/12_2011_117

42. Jayakumar R, Prabaharan M, Kumar P, Nair S, Tamura H (2011) Biomaterials based on chitin and chitosan in wound dressing applications. Biotechnol Adv 29:322–337. https://doi.org/1 0.1016/j.biotechadv.2011.01.005

43. Hirano S, Bash E (2001) wet-spinning and applications of functional fibers based on chitin and chitosan. Macromol Symp 168:21–30. https://doi.org/10.1017/CBO9781107415324.004

44. Wei YC, Hudson SM, Mayer JM (1992) The crosslinking of chitosan fibers. J Polym SciA 30:2187–2193

45. Lee KY, Mooney DJ (2012) Alginate: properties and biomedical applications. Prog Polym Sci 37:106–126. https://doi.org/10.1016/j.progpolymsci.2011.06.003

46. Song R, Xue R, He L, Liu Y, Xiao Q (2008) The structure and properties of chitosan/polyethylene glycol/silica ternary hybrid organic-inorganic films. Chin J Polym Sci 26:621–630

47. Xie H, Zhang S, Li S (2006) Chitin and chitosan dissolved in ionic liquids as reversible sorbents of CO_2. Green Chem 8:630–633. https://doi.org/10.1039/b517297g

48. Han C, Zhang L, Sun J, Shi H, Zhou J, Gao C (2010) Application of collagen-chitosan/fibrin glue asymmetric scaffolds in skin tissue engineering. Biomed & Biotechnol 11:524–530. https://doi.org/10.1631/jzus.B0900400

49. Ma L, Gao C, Mao Z, Zhou J, Shen J (2003) Collagen/chitosan porous scaffolds with improved biostability for skin tissue engineering. Biomater 24:4833–4841. https://doi.org/10.1016/S0 142-9612(03)00374-0

50. Tangsadthakun C, Kanokpanont S, Sanchavanakit N, Banaprasert T, Damrongsakkul S (2006) Properties of collagen/chitosan scaffolds for skin tissue engineering fabrication of collagen/chitosan scaffolds. J Met Mater Miner 16:37–44

51. Wang X, Yan Y, Xiong Z, Lin F, Wu R, Zhang R, Lu Q (2005) Preparation and evaluation of ammonia-treated collagen/chitosan matrices for liver tissue engineering. J Biomed Mater Res B Appl Biomater 75B:91–98. https://doi.org/10.1002/jbm.b.30264

52. Abbas AA, Lee SY, Selvaratnam L, Yusof N, Kamarul T (2008) Porous PVA-chitosan based hydrogel as an extracellular matrix scaffold for cartilage regeneration. Eur Cell Mater 16:50–51
53. Ang TH, Sultana FSA, Hutmacher DW, Wong YS, Fuh JYH, Mo XM, Loh HT, Burdet E, Teoh SH (2002) Fabrication of 3D chitosan—hydroxyapatite scaffolds using a robotic dispensing system. Mater Sci Eng C 20:35–42
54. Breyner NM, Zonari AA, Carvalho JL, Gomide VS, Gomes D, Góes AM (2011) Cartilage tissue engineering using mesenchymal stem cells and 3D chitosan scaffolds—in vitro and in vivo assays. Biomaterials science and engineering. InTech Published, Institute of Biologic Science, Department of Biochemistry and Immunology, Brazil, pp 211–226
55. Iqbal M, Xiaoxue SÆ (2009) A review on biodegradable polymeric materials for bone tissue engineering applications. J Mater Sci 44:5713–5724. https://doi.org/10.1007/s10853-009-3770-7
56. Hussain A, Collins G, Yip D, Cho CH (2012) Functional 3-D cardiac co-culture model using bioactive chitosan nanofiber scaffolds. Biotech Bioeng 110:1–11. https://doi.org/10.1002/bit.24727
57. Martins A, Reis RL, Neves NM (2007) Electrospun nanostructured scaffolds for tissue engineering applications. Nanomedi 2:929–942
58. Zhang T, Wan LQ, Xiong Z, Marsano A, Maidhof R, Park M, Yan Y, Vunjak-novakovic G (2012) Channelled scaffolds for engineering myocardium with mechanical stimulation. J Tissue Eng Regen Med 6:748–756. https://doi.org/10.1002/term
59. Koo S, Ahn SJ, Hao Z, Wang JC, Yim EK (2011) Human corneal keratocyte response to micro- and nano-gratings on chitosan and PDMS. Cell Mol Bioeng 4:399–410. https://doi.org/10.1007/s12195-011-0186-7
60. Draget KI, Smidsrùd PO, Skjåk-brñk PG (2005) Alginates from Algae. In: Polysccharides and polyamides in the food industry. Properties, production and patents. Wiley, Weinheim, pp 1–30
61. Wang L, Li C, Chen Y, Dong S, Chen X, Zhou Y (2013) Poly (lactic-co-glycolic) acid/nanohydroxyapatite scaffold containing chitosan microspheres with adrenomedullin delivery for modulation activity of osteoblasts and vascular endothelial cells. Biomed Res Int 2013:1–13
62. Bhattarai N, Gunn J, Zhang M (2010) Chitosan-based hydrogels for controlled, localized drug delivery. Adv Drug Deliv Rev 62:83–99. https://doi.org/10.1016/j.addr.2009.07.019
63. Patel MP, Patel RR, Patel JK (2010) Chitosan mediated targeted drug delivery system: a review. J Pharm Pharm Sci 13:536–557
64. East G, Qin Y (2003) Wet spinning of chitosan and the acetylation of chitosan fibers. J Appl Polym Sci 50:1773–1779. https://doi.org/10.1002/app.1993.070501013
65. Wan Y, Cao X, Zhang S, Wang S, Wu Q (2008) Fibrous poly(chitosan-g-dl-lactic acid) scaffolds prepared via electro-wet-spinning. Acta Biomater 4:876–886. https://doi.org/10.1016/j.actbio.2008.01.001
66. Doucet BM, Lam A, Griffin L (2012) Neuromuscular electrical stimulation for skeletal muscle function. Yale J Biol Med 85:201–215
67. Hsu M, Wei S, Chang Y, Gung C (2011) Effect of neuromuscular electrical muscle stimulation on energy expenditure in healthy adults. Sensors 11:1932–1942. https://doi.org/10.3390/s110201932
68. Longo U, Loppini M, Berton A, Spiezia F, Maffulli N, Denaro V (2012) Tissue engineered strategies for skeletal muscle injury. Stem Cells Int 2012:175038. https://doi.org/10.1155/2012/175038
69. Meng S, Rouabhia M, Zhang Z, De D, De F, Laval U (2011) Electrical stimulation in tissue regeneration. In: Applied biomedical engineering, pp 37–62
70. Rupp A, Dornseifer U, Fischer A, Schmahl W, Rodenacker K, Uta J, Gais P, Biemer E, Papadopulos N, Matiasek K (2007) Electrophysiologic assessment of sciatic nerve regeneration in the rat: surrounding limb muscles feature strongly in recordings from the gastrocnemius muscle. J Neurosci Methods 166:266–277. https://doi.org/10.1016/j.jneumeth.2007.07.015

71. Wei Z (2014) Nanoscale tunable reduction of graphene oxide for graphene electronics. Science (80)1373:1372–1376. https://doi.org/10.1126/science.1188119
72. Chao Y, Chao EY, Inoue N (2003) Biophysical stimulation of bone fracture repair, regeneration and remodelling. Eur Cell Mater 6:72–85
73. Lafayette W (2003) Criteria for the selection of materials for implanted electrodes. Anal Biomed Eng 31:879–890. https://doi.org/10.1114/1.1581292
74. Green RA, Hassarati RT, Goding JA, Baek S, Lovell NH, Martens PJ, Poole-Warren LA (2012) Conductive hydrogels: mechanically robust hybrids for use as biomaterials. Macromol Biosci 12:494–501. https://doi.org/10.1002/mabi.201100490
75. Kim DOK (2001) High temperature mechanical properties. Platin Metels Rev 45:74–82
76. Lacour SP, Benmerah S, Tarte E, Fitzgerald J, Serra J, McMahon S, Fawcett J, Graudejus O, Yu Z, Morrison B (2010) Flexible and stretchable micro-electrodes for in vitro and in vivo neural interfaces. Med Biol Eng Comput 48:945–954. https://doi.org/10.1007/s11517-010-0644-8
77. Ludwig KA, Uram JD, Yang J, Martin DC, Kipke DR (2006) Chronic neural recordings using silicon microelectrode arrays electrochemically deposited with a poly(3,4-ethylenedioxythiophene) (PEDOT) film. J Neural Eng 3:59–70
78. Humayun MS, Weiland JD, Fujii GY, Greenberg R, Williamson R, Little J, Mech B, Cimmarusti V, Van Boemel G, Dagnelie G, De Juan E (2003) Visual perception in a blind subject with a chronic microelectronic retinal prosthesis. Vision Res 43:2573–2581. https://doi.org/10.1016/S0042-6989(03)00457-7
79. Ben-jacob E, Hanein Y (2008) Carbon nanotube micro-electrodes for neuronal interfacing. J Mater Chem 18:5181–5186. https://doi.org/10.1039/b805878b
80. Wallace G, Moulton S, Kapsa R, Higgins M (2012) Key elements of a medical bionic device. In: Organic bionics, p 240
81. Katsnelson MI (2007) Graphene: carbon in two dimensions. Mater Today 10:20–27. https://doi.org/10.1016/S1369-7021(06)71788-6
82. Jalili R (2012) Wet-spinning of nanostructured fibres. University of Wollongong
83. Li D, Müller M, Gilje S, Kaner RB, Wallace GG (2008) Processable aqueous dispersions of graphene nanosheets. Nat Nanotechnol 3:101–106. https://doi.org/10.1038/nnano.2007.451
84. Cheng H, Hu C, Zhao Y, Qu L (2014) Graphene fiber: a new material platform for unique applications. NPG Asia Mater 6:e113. https://doi.org/10.1038/am.2014.48
85. Xu Z, Gao C (2011) Graphene chiral liquid crystals and macroscopic assembled fibres. Nat Commun 2:571–579. https://doi.org/10.1038/ncomms1583
86. Cong H-P, Ren X-C, Wang P, Yu S-H (2012) Wet-spinning assembly of continuous, neat, and macroscopic graphene fibers. Sci Rep 2:613. https://doi.org/10.1038/srep00613
87. Jalili R, Aboutalebi H, Esrafilzadeh D, Shepherd R, Chen J, Aminorroaya-yamini S, Konstantinov K, Minett A, Razal J, Wallace G (2013) Scalable one-step wet-spinning of graphene fibers and yarns from liquid crystalline dispersions of graphene oxide: towards multifunctional textiles. Adv Funct Mater 23:5345–5354. https://doi.org/10.1002/adfm.201300765
88. Fan X, Peng W, Li Y, Li X, Wang S, Zhang G, Zhang F (2008) Deoxygenation of exfoliated graphite oxide under alkaline conditions: a green route to graphene preparation. Adv Mater 20:4490–4493. https://doi.org/10.1002/adma.200801306
89. Rourke J, Pandey P, Moore J, Bates M, Kinloch I, Young R, Wilson NR (2011) The real graphene oxide revealed: stripping the oxidative debris from the graphene-like sheets. Angew Chem Int Ed 50:3173–3177. https://doi.org/10.1002/anie.201007520
90. Shirakawa H, Louis EJ, MacDiarmid AG, Chiang CK, Heeger AJ (1977) Synthesis of electrically conducting organic polymers: halogen derivatives of polyacetylene, (CH)x. JCS Chem Comm 16:578–580
91. Street TC, Clarke GB (1981) Conducting polymers: a review of recent work. IBM J Res Dev 25:51–57
92. Macdiarmid AG (2001) Synthetic metals: a novel role for organic polymers (nobel lecture). Angew Chem Int Ed 40:2581–2590

93. Chandrasekhar P (1999) Conducting polymers: fundamentals and applications. Kluwer Academic Publishers
94. Diaz AF, Kanazawa KK, Gardini GP (1979) Electrochemical polymerization of pyrrole. JCS Chem Commun 635–636. https://doi.org/10.1039/c39790000635
95. Lin Y, Wallace GG (1994) Factors influencing electrochemical release of 2,6-anthraquinone disulphonic acid from polypyrrole. J Control Release 30:137–142
96. Spinks GM, Xi B, Truong V, Wallace GG (2005) Actuation behaviour of layered composites of polyaniline, carbon nanotubes and polypyrrole. Synth Metels 151:85–91. https://doi.org/1 0.1016/j.synthmet.2005.03.006
97. Wallace GG, Kane-maguire LAP (2002) Manipulating and monitoring biomolecular interactions with conducting electroactive polymers. Adv Mater 14:953–960
98. Wang J, Too CO, Zhou D, Wallace GG (2005) Novel electrode substrates for rechargeable lithium/polypyrrole batteries. J Power Sources 140:162–167. https://doi.org/10.1016/j.jpows our.2004.08.040
99. Barisci JN, Stella R, Spinks GM, Wallace GG (2001) Study of the surface potential and photovoltage of conducting polymers using electric force microscopy. Synth Metels 124:407–414
100. Kane-Maguire LA, Norris ID, Wallace GG (1999) Properties of chid polyaniline in various oxidation states. Synth Metels 101:817–818
101. Kane-maguire LAP, Macdiarmid AG, Norris ID, Wallace GG (1999) Facile preparation of optically active polyanilines via the in situ chemical oxidative polymerisation of aniline. Synth Metels 106:171–176
102. Liu C, Lin C, Kuo C, Lin S, Chen W (2011) Conjugated rod—coil block copolymers: synthesis, morphology, photophysical properties, and stimuli-responsive applications. Prog Polym Sci 36:603–637. https://doi.org/10.1016/j.progpolymsci.2010.07.008
103. Wang C, Ballantyne A, Hall S, Too C, Officer D, Wallace G (2006) Functionalized polythiophene-coated textile: a new anode material for a flexible battery. J Power Sources 156:610–614. https://doi.org/10.1016/j.jpowsour.2005.06.020
104. Li C, Bai H, Shi G (2009) Conducting polymer nanomaterials: electrosynthesis and applications. Chem Soc Rev 38:2149–2496. https://doi.org/10.1039/b816681c
105. Wan M (2009) Some issues related to polyaniline micro-/nanostructures. Macromol Rapid Commun 30:963–975. https://doi.org/10.1002/marc.200800817
106. Heeger AJ (2002) Semiconducting and metallic polymers: the fourth generation of polymeric materials. Synth Metels 125:23–42
107. Laslau C, Zujovic Z, Travas-sejdic J (2010) Theories of polyaniline nanostructure self-assembly: towards an expanded, comprehensive multi-layer theory (MLT). Prog Polym Sci 35:1403–1419. https://doi.org/10.1016/j.progpolymsci.2010.08.002
108. Stejskal J, Sapurina I, Trchová M (2010) Polyaniline nanostructures and the role of aniline oligomers in their formation. Prog Polym Sci 35:1420–1481. https://doi.org/10.1016/j.progp olymsci.2010.07.006
109. Semire B, Odunola OA (2011) Semiempirical and density functional theory study on structure of fluoromethylfuran oligomers. Aust J Bas Appl Sci 5:354–359
110. Tamer U, Kanbeş Ç, Torul H, Ertaş N (2011) Preparation, characterization and electrical properties of polyaniline nanofibers containing sulfonated cyclodextrin group. React Funct Polym 71:933–937. https://doi.org/10.1016/j.reactfunctpolym.2011.06.002
111. Gupta N, Sharma S, Mir I, Kumar D (2006) Advances in sensors based on conducting polymers. J Sci Ind Res 65:549–557
112. Xia L, Wei Z, Wan M (2010) Conducting polymer nanostructures and their application in biosensors. J Colloid Interface Sci 341:1–11. https://doi.org/10.1016/j.jcis.2009.09.029
113. Guimard NK, Gomez N, Schmidt CE (2007) Conducting polymers in biomedical engineering. Prog Polym Sci 32(32):876–921. https://doi.org/10.1016/j.progpolymsci.2007.05.012
114. Huang ZB, Yin GF, Liao XM, Gu JW (2014) Conducting polypyrrole in tissue engineering applications. Front Mater Sci 8:39–45. https://doi.org/10.1007/s11706-014-0238-8

115. Jager EWH, Immerstrand C, Magnusson K, Inganas O, Lundstrom I (2000) Biomedical applications of polypyrrole microactuators : from single-cell clinic to microrobots. In: Annual international IEEE-EMBS special topic conference on microtechnologies in medicine & biology, pp 58–61

116. Min Y, Yang Y, Poojari Y, Liu Y, Wu J, Hansford DJ, Epstein AJ (2013) Sulfonated Polyaniline-based organic electrodes for controlled electrical stimulation of human osteosarcoma cells. Biomacromolecules 14:1727–1731

117. Quigley BAF, Razal JM, Thompson BC, Moulton SE, Kita M, Kennedy EL, Clark GM, Wallace GG, Kapsa RMI (2009) A Conducting-polymer platform with biodegradable fibers for stimulation and guidance of axonal growth. Adv Mater 21:1–5. https://doi.org/10.1002/a dma.200901165

118. Lu X, Zhang W, Wang C, Wen T-C, Wei Y (2011) One-dimensional conducting polymer nanocomposites: synthesis, properties and applications. Prog Polym Sci 36:671–712. https:// doi.org/10.1016/j.progpolymsci.2010.07.010

119. Foroughi J, Spinks G, Wallace G (2011) Chemical High strain electromechanical actuators based on electrodeposited polypyrrole doped with di-(2-ethylhexyl) sulfosuccinate. Sens Actuators B 155:278–284. https://doi.org/10.1016/j.snb.2010.12.035

120. Smela BE (2003) Conjugated polymer actuators for biomedical applications. Adv Mater 15:481–494

121. Perepichka IF, Besbes M, Levillain E, Salle M, Roncali J (2002) Hydrophilic oligo (oxyethylene)-derivatized optoelectroelectrochemical properties and solid-state chromism. Chem Mater 14:449–457

122. Esrafilzadeh D, Razal J, Moulton S, Stewart E, Wallace G (2013) Multifunctional conducting fibres with electrically controlled release of ciprofloxacin. J Control Release 169:313–320

123. Seyedin S, Razal JM, Innis PC, Jeiranikhameneh A, Beirne S, Wallace GG (2015) Knitted strain sensor textiles of highly conductive all-polymeric fibers. ACS Appl Mater Interfaces 7:21150–21158. https://doi.org/10.1021/acsami.5b04892

124. Esfandiari A (2008) PPy covered cellulosic and protein fibres using novel covering methods to improve the electrical property. World Appl Sci J 3:470–475

125. Wallace GG, Spinks G, Maxwell Kane-Maguire LA, Teasdale, PR (2009) Conductive electroactive polymers: Intelligent polymer systems. CRC Press, Boca Raton, United States

126. Weng B, Shepherd RL, Crowley K, Killard AJ, Wallace GG (2010) Printing conducting polymers. Analyst 135:2779–2789. https://doi.org/10.1039/c0an00302f

127. Kipphan HH (2001) Handbook of print media. Springer Science & Business Media

128. Earls A, Baya V (2014) The road ahead for 3-D printers. Disruptive Manuf Eff 3D Print 14

129. Gomes TC, Constantino CJL, Lopes EM, Job AE, Alves N (2012) Thermal inkjet printing of polyaniline on paper. Thin Solid Films 520:7200–7204

130. Kulkarni MV, Apte SK, Naik SD, Ambekar JD, Kale BB (2013) Ink-jet printed conducting polyaniline based flexible humidity sensor. Sens Actuators B 178:140–143. https://doi.org/1 0.1016/j.snb.2012.12.046

131. Mabrook MF, Pearson C, Petty MC (2006) Inkjet-printed polypyrrole thin films for vapour sensing. Sens Actuators B 115:547–551. https://doi.org/10.1016/j.snb.2005.10.019

132. Weng B, Morrin A, Shepherd R, Crowley K, Killard AJ, Innis PC, Wallace GG (2014) Wholly printed polypyrrole nanoparticle-based biosensors on fl exible substrate. J Mater Chem B 2:793–799. https://doi.org/10.1039/c3tb21378a

133. Zergioti I, Makrygianni M, Dimitrakis P, Normand P, Chatzandroulis S (2011) Laser printing of polythiophene for organic electronics. Appl Surf Sci 257:5148–5151. https://doi.org/10.1 016/j.apsusc.2010.10.145

134. Fischer JE, Tang X, Scherr EM, Cajipe VB, MacDiarmid AG (1991) Polyaniline fibers and films: stretch-induced orientation and crystallization, morphology, and the nature of the amorphous phase. Synth Metels 43:661–664

135. Jannakoudakis AD, Jannakoudakis PD, Pagalos N, Theodoridou E (1993) Electro-oxidation of aniline and electrochemical behaviour of the produced polyaniline film on carbon-fibre electrodes in aqueous methanolic solutions. Electrochim Acta 38:1559–1566

136. Tzou KT, Gregory RV (1995) Improved solution stability and spinnability of concentrated polyaniline solutions using N,N′-dimethyl propylene urea as the spin bath solvent. Synth Metels 69:109–112

137. Mattes BR, Wang HL, Yang D (1997) Formation of conductive polyaniline fibers drived from highly concentrated emeraldine base solutions. Synth Metels 84:45–49

138. Unni SM, Dhavale VM, Pillai VK, Kurungot S (2010) High Pt utilization electrodes for polymer electrolyte membrane fuel cells by dispersing Pt particles formed by a preprecipitation method on carbon "polished" with polypyrrole. J Phys Chem C 114:14654–14661

139. Dadras MA, Entezami A (1993) New synthesis method of polythiophenes. Iran Polym J 3:2–12

140. Roncali J (1992) Conjugated poiy(th1ophenes): synthesis, functionalizatlon, and applications. Chem Rev 92:711–738

141. Groenendaal BL, Jonas F, Freitag D, Pielartzik H, Reynolds JR (2000) Poly(3,4-ethylenedioxythiophene) and its derivatives: past, present, and future. Adv Mater 12:481–494

142. Zhang X, Macdiarmid AG, Manohar SK (2005) Chemical synthesis of PEDOT nanofibers. Chem Commun 12:5328–5330. https://doi.org/10.1039/b511290g

143. Åkerfeldt M (2015) Electrically conductive textile coatings with PEDOT: PSS. University of Boras

144. Environ E, Alemu D, Wei H, Ho K, Chu C (2012) Environmental Science Highly conductive PEDOT: PSS electrode by simple film treatment with methanol for ITO-free polymer solar cells. Energy Environ Sci 5:9662–9671. https://doi.org/10.1039/c2ee22595f

145. Guo X, Liu X, Lin F, Li H, Fan Y, Zhang N (2015) Highly Conductive transparent organic electrodes with multilayer structures for rigid and flexible optoelectronics. Sci Rep 5:1–9. https://doi.org/10.1038/srep10569

146. Islam MM, Chidembo AT, Aboutalebi SH, Cardillo D, Liu HK, Al E (2014) Liquid crystalline graphene oxide/PEDOT: PSS self-assembled 3D architecture for binder-free supercapacitor electrodes. Front Mater Sci 2:1–21

147. Kim BH, Park DH, Joo J, Yu SG, Lee SH (2005) Synthesis, characteristics, and field emission of doped and de-doped polypyrrole, polyaniline, poly(3,4-ethylenedioxythiophene) nanotubes and nanowires. Synth Metels 150:279–284. https://doi.org/10.1016/j.synthmet.2005.02.012

148. Baik W, Luan W, Zhao RH, Koo S, Kim K (2009) Synthesis of highly conductive poly(3,4-ethylenedioxythiophene) fiber by simple chemical polymerization. Synth Metels 159:1244–1246. https://doi.org/10.1016/j.synthmet.2009.02.044

149. Okuzaki H, Ishihara M (2003) Spinning and characterization of conducting microfibers. Macromol Rapid Commun 24:261–264

150. Okuzaki H, Harashina Y, Yan HH (2009) Highly conductive PEDOT/PSS microfibers fabricated by wet-spinning and dip-treatment in ethylene glycol. Eur Polym J 45:256–261. https://doi.org/10.1016/j.eurpolymj.2008.10.027

151. Jalili R, Razal JM, Innis PC, Wallace GG (2011) One-Step wet-spinning process of poly (3, 4-ethylenedioxy- thiophene): poly (styrenesulfonate) fibers and the origin of higher electrical conductivity. Adv Func Mater 21:3363–3370. https://doi.org/10.1002/adfm.201100785

152. Han D, Lee HJ, Park S (2005) Electrochemistry of conductive polymers XXXV: electrical and morphological characteristics of polypyrrole films prepared in aqueous media studied by current sensing atomic force microscopy. Electrochim Acta 50:3085–3092. https://doi.org/10.1016/j.electacta.2004.10.085

153. Vernitskaya TV, Efimov ON (1997) Polypyrrole: a conducting polymer; its synthesis, properties and applications. Russ Chem Rev 443:443–457

154. Ateh D, Navsaria H, Vadgama P (2006) Polypyrrole-based conducting polymers and interactions with biological tissues. J R Soc Interface 3:741–752. https://doi.org/10.1098/rsif.2006.0141

155. Virji S, Huang J, Kaner RB, Weiller BH (2004) Polyaniline nanofiber gas sensors: examination of response mechanisms. Nano Lett 4:491–496. https://doi.org/10.1021/nl035122eCCC

156. Wu J, Pawliszyn J (2001) Preparation and applications of polypyrrole films in solid-phase microextraction. J Chromatogr A 909:37–52

157. Foroughi J (2009) Development of novel nanostructured conducting polypyrrole fibres. University of Wollongong
158. Kim D, Kim YD (2007) Electrorheological properties of polypyrrole and its composite ER fluids. J Ind Eng Chem 13:879–894
159. Cui CJ, Wu GM, Yang HY, She SF, Shen J, Zhou B, Zhang ZH (2010) A new high-performance cathode material for rechargeable lithium-ion batteries: polypyrrole/vanadium oxide nanotubes. Electrochim Acta 55:8870–8875. https://doi.org/10.1016/j.electacta.2010.07.087
160. Kakuda S, Momma T, Osaka T (1995) Ambient-temperature, rechargeable, all-solid lithium/polypyrrole polymer battery. J Electrochem Soc 142:1–2
161. Li X, Hao X, Yu H, Na H (2008) Fabrication of Polyacrylonitrile/polypyrrole (PAN/Ppy) composite nanofibres and nanospheres with core—shell structures by electrospinning. Mater Lett 62:1155–1158. https://doi.org/10.1016/j.matlet.2007.08.003
162. Saville P (2005) Polypyrrole, formation and use
163. Foroughi J, Spinks GM, Wallace GG, Whitten PG (2008) Production of polypyrrole fibres by wet spinning. Synth Metals 158:104–107. https://doi.org/10.1016/j.synthmet.2007.12.008
164. Rowley NM, Mortimer RJ (2002) New electrochromic materials. Sci Prog 85:243–262
165. Mccullough LA, Dufour B, Matyjaszewski K (2009) Polyaniline and polypyrrole templated on self-assembled acidic block copolymers. Macromolecules 42:8129–8137. https://doi.org/10.1021/ma901560k
166. Li Y, Cheng XY, Leung MY, Tsang J, Tao XM, Yuen MCW (2005) A flexible strain sensor from polypyrrole-coated fabrics. Synth Metals 155:89–94. https://doi.org/10.1016/j.synthmet.2005.06.008
167. Xing S, Zhao G (2007) Morphology, structure, and conductivity of polypyrrole prepared in the presence of mixed surfactants in aqueous solutions. J Appl Polym Sci 104:1987–1996. https://doi.org/10.1002/app
168. Grunden B, Iroh JO (1995) Formation of graphite fibre polypyrrole coatings by aqueous electrochemical polymerization. Polym J 36:559–563
169. Flores O, Romo-Uribe A, Romero-Guzman ME, Gonzalez AE, Cruz-Silva R, Campillo B (2009) Mechanical properties and fracture behavior of polypropylene reinforced with polyaniline-grafted short glass fibers. J Appl Polym Sci 112:934–941. https://doi.org/10.1002/app
170. Granato F, Bianco A, Bertarelli C, Zerbi G (2009) Composite polyamide 6/polypyrrole conductive nanofibers. Macromol Rapid Commun 30:453–458. https://doi.org/10.1002/marc.200800623
171. Nair S, Natarajan S, Kim S (2005) Fabrication of electrically conducting polypyrrole-poly (ethylene oxide) composite nanofibers. Macromol Rapid Commun 26:1599–1603. https://doi.org/10.1002/marc.200500457
172. Wang H, Leaukosol N, He Z (2013) Microstructure, distribution and properties of conductive polypyrrole/cellulose fiber composites. Cellulose 20:1587–1601. https://doi.org/10.1007/s10570-013-9945-z
173. Kim CY, Lee JY, Kim DY (1998) Soluble, electroconductive polypyrrole and method for preparing the same. US5795953
174. Lee GJ, Lee SH, Ahn KS, Kim KH (2002) Synthesis and characterization of soluble polypyrrole with improved electrical conductivity. J Appl Polym Sci 84:2583–2590. https://doi.org/10.1002/app.10281
175. Oh EJ, Jang KS (2001) Synthesis and characterization of high molecular weight, highly soluble polypyrrole in organic solvents. Synth Metals 119:109–110
176. Qi Z, Pickup PG (1997) Size control of polypyrrole particles. Chem Mater 9:2934–2939
177. Li BS, Macosko CW, White HS (1993) Electrochemical processing of electrically conductive polymer fibers. Adv Mater 5:575–576
178. Cho JW, Jung H (1997) Electrically conducting high-strength aramid composite fibres prepared by vapour-phase polymerization of pyrrole. J Mater Sci 32:5371–5376
179. Gholivand MB, Abolghasemi MM, Fattahpour P (2011) Polypyrrole/hexagonally ordered silica nanocomposite as a novel fiber coating for solid-phase microextraction. Anal Chim Acta 704:174–179. https://doi.org/10.1016/j.aca.2011.07.045

180. Maziz A, Khaldi A, Persson N, Jager EWH (2015) Soft linear electroactive polymer actuators based on polypyrrole. In: Proceedings of SPIE, pp 1–6

181. Xu C, Wang P, Bi X (1995) Continuous vapor phase polymerization of pyrrole. I. Electrically conductive composite fiber of polypyrrole with poly(p-phenylene terephthalamide). J Appl Polym Sci 58:2155–2159

182. Chronakis IS, Grapenson S, Jakob A (2006) Conductive polypyrrole nanofibers via electrospinning: electrical and morphological properties. Polym J 47:1597–1603. https://doi.org/10.1016/j.polymer.2006.01.032

183. Srivastava Y, Loscertales I, Marquez M, Thorsen T (2007) Electrospinning of hollow and core/sheath nanofibers using a microfluidic manifold. Microfluid Nanofluid 4:245–250. https://doi.org/10.1007/s10404-007-0177-0

184. Sen S, Davis FJ, Mitchell GR, Robinson E (2009) Conducting nanofibres produced by electrospinning. J Phys 183:12–20. https://doi.org/10.1088/1742-6596/183/1/012020

185. Hamilton S, Hepher MJ, Sommerville J (2005) Polypyrrole materials for detection and discrimination of volatile organic compounds. Sens Actuators B 107:424–432. https://doi.org/10.1016/j.snb.2004.11.001

186. Maity S, Chatterjee A (2015) Textile/polypyrrole composites for sensory applications. J Compos 2015:1–6

187. Geiger B, Bershadsky A, Pankov R, Yamada KM, Correspondence BG (2001) Transmembrane extracellular matrix—cytoskeleton crosstalk. Nat Rev 2:793–805. https://doi.org/10.1038/35099066

188. Farra N (2008) Development and characterization of conductive polyaniline fibre actuators. University of Toronto

189. Agarwal S, Wendorff JH, Greiner A (2008) Use of electrospinning technique for biomedical applications. Polymer (Guildf) 49:5603–5621. https://doi.org/10.1016/j.polymer.2008.09.014

190. Reneker DH, Chun I (1996) Nanometre diameter fibres of polymer, produced by electrospinning. Nanotechnology 7:216–223

191. Huang ZM, Zhang Y-Z, Kotaki M, Ramakrishna S, Huang Z-M, Zhang Y-Z, Kotaki SR (2003) A review on polymer nanofibers by electrospinning and their applications in nanocomposites. Compos Sci Technol 63:2223–2253

192. Bhattarai P, Thapa KB, Basnet RB, Sharma Saurav (2014) Electrospinning: how to produce nanofibers using most inexpensive technique? An insight into the real challenges of electrospinning such nanofibers and its application areas. IJBAR 5:79–80. https://doi.org/10.7439/ijbar

193. Ziebicki A (2010) Fundamentals of fibre formation. Wiley, London

194. Srivastava Y, Marquez M, Thorsen T (2009) Microfluidic electrospinning of biphasic nanofibers with Janus morphology. Biomicrofluidics 3:12801. https://doi.org/10.1063/1.3009288

195. Mirabedini A, Foroughi J, Wallace GGGG (2016) Developments in conducting polymer fibres: from established spinning methods toward advanced applications. RSC Adv 6:44687–44716. https://doi.org/10.1039/C6RA05626A

196. Elahi F, Lu W, Guoping G, Khan F (2013) Core-shell fibers for biomedical applications—a review. J Bioeng Biomed Sci 3:1–14. https://doi.org/10.4172/2155-9538.1000121

197. Khan SN (2007) Electrospinning polymer nanofibers-electrical and optical characterization. Ohio University

198. Dersch R, Liu T, Schaper AK, Greiner A, Wendorff JH (2003) Electrospun nanofibers: internal structure and intrinsic orientation. J Polym Sci A 41:545–553. https://doi.org/10.1002/pola.10609

199. Dong B, Arnoult O, Smith ME, Wnek GE (2009) Electrospinning of collagen nanofiber scaffolds from benign solvents. Macromol Rapid Commun 30:539–542. https://doi.org/10.1002/marc.200800634

200. Greiner A, Wendorff JH, Yarin AL, Zussman E (2006) Biohybrid nanosystems with polymer nanofibers and nanotubes. Appl Microbiol Biotechnol 71:387–393. https://doi.org/10.1007/s00253-006-0356-z

201. Jayakumar R, Prabaharan M, Kumar PTS (1990) Novel chitin and chitosan materials in wound dressing. Biomedical engineering, trends in materials science. InTech Amrita Centre for Nanosciences and Molecular Medicine, India, pp 3–25

202. Jeong SI, Ph D, Krebs MD, Bonino CA, Samorezov JE, Khan SA, Alsberg E (2011) In situ polyelectrolyte complexation for use as tissue engineering scaffolds. TISSUE Eng Part A 17. https://doi.org/10.1089/ten.tea.2010.0086

203. Lee Y-S, Livingston Arinzeh T (2011) Electrospun nanofibrous materials for neural tissue engineering. Polymers (Basel) 3:413–426. https://doi.org/10.3390/polym3010413

204. Wang J, Huang X, Xiao J, Li N, Yu W, Wang W, Xie W, Ma X, Teng Y (2010) Spray-spinning: a novel method for making alginate/chitosan fibrous scaffold. J Mater Sci 21:497–506. https://doi.org/10.1007/s10856-009-3867-1

205. Xuejun Xin MH (2007) Continuing differentiation of human mesenchymal stem cells and osteogenic lineage in electrospun PLGA nanofiber scaffold. Biomaterials 28:316–325. https://doi.org/10.1016/j.biomaterials.2006.08.042

206. Zong X, Bien H, Chung CY, Yin L, Fang D, Hsiao BS, Chu B, Entcheva E (2005) Electrospun fine-textured scaffolds for heart tissue constructs. Biomaterials 26:5330–5338. https://doi.org/10.1016/j.biomaterials.2005.01.052

207. Ghasemi-Mobarakeh L, Prabhakaran MP, Morshed M, Nasr-Esfahani MH, Ramakrishna S (2009) Electrical stimulation of nerve cells using conductive nanofibrous scaffolds for nerve tissue engineering. Tissue Eng: A 15:3605–3619. https://doi.org/10.1089/ten.tea.2008.0689

208. Xu X, Chen X, Ma P, Wang X, Jing X (2008) The release behavior of doxorubicin hydrochloride from medicated fibers prepared by emulsion-electrospinning. Eur J Pharm Biopharm 70:165–170. https://doi.org/10.1016/j.ejpb.2008.03.010

209. Xu X, Chen X, Wang Z, Jing X (2009) Ultrafine PEG-PLA fibers loaded with both paclitaxel and doxorubicin hydrochloride and their in vitro cytotoxicity. Eur J Pharm Biopharm 72:18–25. https://doi.org/10.1016/j.ejpb.2008.10.015

210. Ho Y-C, Huang F-M, Chang Y-C (2007) Cytotoxicity of formaldehyde on human osteoblastic cells is related to intracellular glutathione levels. J Biomed Mater Res B Appl Biomater 83:340–344. https://doi.org/10.1002/jbmb

211. Huang ZM, He CL, Yang A, Zhang Y, Han XJ, Yin J, Wu Q (2006) Encapsulating drugs in biodegradable ultrafine fibers through co-axial electrospinning. J Biomed Mater Res Part A 77:169–179. https://doi.org/10.1002/jbm.a.30564

212. Kim K, Luu YK, Chang C, Fang D, Hsiao BS, Chu B, Hadjiargyrou M (2004) Incorporation and controlled release of a hydrophilic antibiotic using poly(lactide-co-glycolide)-based electrospun nanofibrous scaffolds. J Control Release 98:47–56. https://doi.org/10.1016/j.jconrel.2004.04.009

213. Xie J, Wang CH (2006) Electrospun micro- and nanofibers for sustained delivery of paclitaxel to treat C6 glioma in vitro. Pharm Res 23:1817–1826. https://doi.org/10.1007/s11095-006-9036-z

214. Xu X, Chen X, Xu X, Lu T, Wang X, Yang L, Jing X (2006) BCNU-loaded PEG-PLLA ultrafine fibers and their in vitro antitumor activity against Glioma C6 cells. J Control Release 114:307–316. https://doi.org/10.1016/j.jconrel.2006.05.031

215. Chew SY, Wen J, Yim EKF, Leong KW (2005) Sustained release of proteins from electrospun biodegradable fibers. Biomacromolecules 6:2017–2024. https://doi.org/10.1021/bm0501149

216. Jiang H, Hu Y, Li Y, Zhao P, Zhu K, Chen W (2005) A facile technique to prepare biodegradable coaxial electrospun nanofibers for controlled release of bioactive agents. J Control Release 108:237–243. https://doi.org/10.1016/j.jconrel.2005.08.006

217. Nie H, Wang CH (2007) Fabrication and characterization of PLGA/HAp composite scaffolds for delivery of BMP-2 plasmid DNA. J Control Release 120:111–121. https://doi.org/10.1016/j.jconrel.2007.03.018

218. Repanas A, Wolkers W, Müller M, Gryshkov O, Glasmacher B (2015) Pcl/Peg electrospun fibers as drug carriers for the controlled delivery of dipyridamole. J Silico Vitr Pharmacol 1:1–10

219. Maleknia L, Rezazadeh Majdi Z (2014) Electrospinning of gelatin nanofiber for biomedical application. Orient J Chem 30:2043–2048. https://doi.org/10.13005/ojc/300470

220. Khalil KA, Fouad H, Elsarnagawy T, Almajhdi FN (2013) Preparation and characterization of electrospun PLGA/silver composite nanofibers for biomedical applications. Int J Electrochem Sci 8:3483–3493

221. Talebian S, Mehrali MM, Mohan S, Balaji raghavendran HR, Mehrali MM, Khanlou HM, Kamarul T, Afifi AM, Abass AA (2014) Chitosan (PEO)/bioactive glass hybrid nanofibers for bone tissue engineering. RSC Adv 4:49144–49152. https://doi.org/10.1039/C4RA06761D

222. Krogstad EA, Woodrow KA (2014) Manufacturing scale-up of electrospun poly(vinyl alcohol) fibers containing tenofovir for vaginal drug delivery. Int J Pharm 475:282–291. https://doi.org/10.1016/j.ijpharm.2014.08.039

223. Hu C, Gong RH, Zhou FL (2015) Electrospun sodium alginate/polyethylene oxide fibers and nanocoated yarns. Int J Polym Sci 2015:1–12. https://doi.org/10.1155/2015/126041

224. Khil M-S, Cha D-I, Kim H-Y, Kim I-S, Bhattarai N (2003) Electrospun nanofibrous polyurethane membrane as wound dressing. J Biomed Mater Res B Appl Biomater 67:675–679. https://doi.org/10.1002/jbm.b.10058

225. Balaji Raghavendran HRB, Puvaneswary S, Talebian S, Raman Murali M, Vasudevaraj Naveen S, Krishnamurithy G, McKean R, Kamarul T (2014) A comparative study on in vitro osteogenic priming potential of electron spun scaffold PLLA/HA/Col, PLLA/HA, and PLLA/Col for tissue engineering application. PLoS ONE 9:e104389. https://doi.org/10.1371/journal.pone.0104389

226. Ali S, Khatri Z, Oh KW, Kim IS, Kim SH (2014) Preparation and characterization of hybrid polycaprolactone/cellulose ultrafine fibers via electrospinning. Macromol Res 22:562–568

227. Ren X, Akdag A, Zhu C, Kou L, Worley SD, Huang TS (2009) Electrospun polyacrylonitrile nanofibrous biomaterials. J Biomed Mater Res A 91:385–390. https://doi.org/10.1002/jbm.a.32260

228. Hilal Algan A, Pekel-Bayramgil N, Turhan F, Altanlar N (2015) Ofloxacin loaded electrospun fibers for ocular drug delivery. Curr Drug Deliv

229. Venugopal J, Ma LL, Yong T, Ramakrishna S (2005) In vitro study of smooth muscle cells on polycaprolactone and collagen nanofibrous matrices. Cell Biol Int 29:861–867. https://doi.org/10.1016/j.cellbi.2005.03.026

230. Shin M, Ishii O, Sueda T, Vacanti JP (2004) Contractile cardiac grafts using a novel nanofibrous mesh. Biomaterials 25:3717–3723. https://doi.org/10.1016/j.biomaterials.2003.10.055

231. Li W, Laurencin CT, Caterson EJ, Tuan RS, Ko FK (2002) Electrospun nanofibrous structure: a novel scaffold for tissue engineering. J Biomed Mater Res B Appl Biomater 60:613–621. https://doi.org/10.1002/jbm.10167

232. Rujitanaroj PO, Pimpha N, Supaphol P (2008) Wound-dressing materials with antibacterial activity from electrospun gelatin fiber mats containing silver nanoparticles. Polymer (Guildf) 49:4723–4732. https://doi.org/10.1016/j.polymer.2008.08.021

233. Moncrieff RW (1970) Man-Made Fibres. Wiley, Illustrate

234. Gupta MN, Sengupta AK, Kothari V (1997) Manufactured fibre technology. Springer Science & Business Media

235. Huang T, Marshall LR, Armantrout JE, Yembrick S, Dunn WH, Oconnor JM, Mueller T, Avgousti M, Wetzel MD (2012) Production of nanofibers by melt spinning. 3–6

236. Jia J, Yao D, Wang Y (2014) Melt spinning of continuous filaments by cold air attenuation melt spinning of continuous filaments by cold air. Text Res J 84:604–613

237. Woodings C (2001) Regenerated cellulose fibres, illustrate. CRC Press

238. Dogine K (1970) Formation of fibers and development their structure: wet spinning and dry spinning. The society of fiber science and technology

239. Kang E, Jeong GS, Choi YY, Lee KH, Khademhosseini A, Lee S-H (2011) Digitally tunable physicochemical coding of material composition and topography in continuous microfibres. Nat Mater 10:877–883. http://www.nature.com/nmat/journal/v10/n11/abs/nmat3108.html#supplementary-information

240. Liu H, Xu W, Zou H, Ke G, Li W, Ouyang C (2008) Feasibility of wet spinning of silk-inspired polyurethane elastic biofiber. Mater Lett 62:1949–1952. https://doi.org/10.1016/j.matlet.20 07.10.061

241. Puppi D, Piras AM, Chiellini F, Chiellini E, Martins A, Leonor IB, Neves N, Reis R (2011) Optimized electro- and wet-spinning techniques for the production of polymeric fibrous scaffolds loaded with bisphosphonate and hydroxyapatite. J Tissue Eng Regenerative Med 5:253–263. https://doi.org/10.1002/term.310

242. Tuzlakoglu K, Pashkuleva I, Rodrigues MT, Gomes ME, Van Lenthe GH, Muller R, Reis RL (2010) A new route to produce starch-based fiber mesh scaffolds by wet spinning and subsequent surface modification as a way to improve cell attachment and proliferation. J Biomed Mater Res A 92:369–377. https://doi.org/10.1002/jbm.a.32358

243. Caves JM, Cui W, Wen J, Kumar VA, Haller CA, Chaikof EL (2011) Elastin-like protein matrix reinforced with collagen microfibers for soft tissue repair. Biomaterials 32:5371–5379. https://doi.org/10.1016/j.biomaterials.2011.04.009

244. Puppi D, Mota C, Gazzarri M, Dinucci D, Gloria A, Myrzabekova M, Ambrosio L, Chiellini F (2012) Additive manufacturing of wet-spun polymeric scaffolds for bone tissue engineering. Biomed Microdevices 14:1115–1127. https://doi.org/10.1007/s10544-012-9677-0

245. Cornwell KG, Pins GD (2010) Enhanced proliferation and migration of fibroblasts on the surface of fibroblast growth factor-2-loaded fibrin microthreads. Tissue Eng: A 16:3669–3677. https://doi.org/10.1089/ten.TEA.2009.0600

246. Hwang CM, Khademhosseini A, Park Y, Sun K, Lee SH (2008) Microfluidic chip-based fabrication of PLGA microfiber scaffolds for tissue engineering. Langmuir 24:6845–6851. https://doi.org/10.1021/la800253b

247. Lu HH, Cooper JA, Manuel S, Freeman JW, Attawia MA, Ko FK, Laurencin CT (2005) Anterior cruciate ligament regeneration using braided biodegradable scaffolds: in vitro optimization studies. Biomaterials 26:4805–4816. https://doi.org/10.1016/j.biomaterials.2004.11.050

248. Razal JM, Gilmore KJ, Wallace GG (2008) Carbon nanotube biofiber formation in a polymer-free coagulation bath. Adv Funct Mater 18:61–66. https://doi.org/10.1002/adfm.200700822

249. Lavin DM, Harrison MW, Tee LY, Wei KA, Mathiowitz E (2012) A novel wet extrusion technique to fabricate self-assembled microfiber scaffolds for controlled drug delivery. J Biomed Mater Res A 100 A:2793–2802. https://doi.org/10.1002/jbm.a.34217

250. Cronin EM, Thurmond FA, Williams RS, Wright WE, Nelson KD, Garner HR (2004) Protein-coated poly(L-lactic acid) fibers provide a substrate for differentiation of human skeletal muscle cells. J Biomed Mater Res A 69:373–381. https://doi.org/10.1002/jbm.a.30009

251. Yilgor P, Tuzlakoglu K, Reis RL, Hasirci N, Hasirci V (2009) Incorporation of a sequential BMP-2/BMP-7 delivery system into chitosan-based scaffolds for bone tissue engineering. Biomaterials 30:3551–3559. https://doi.org/10.1016/j.biomaterials.2009.03.024

252. Jung MR, Shim IK, Kim ES, Park YJ, Il Yang Y, Lee SK, Lee SJ (2011) Controlled release of cell-permeable gene complex from poly(L-lactide) scaffold for enhanced stem cell tissue engineering. J Control Release 152:294–302. https://doi.org/10.1016/j.jconrel.2011.03.002

253. Chiang CY, Mello CM, Gu J, Silva ECCM, Van Vliet KJ, Belcher AM (2007) Weaving genetically engineered functionality into mechanically robust virus fibers. Adv Mater 19:826–832. https://doi.org/10.1002/adma.200602262

254. Palakurthi NK, Correa ZM, Augsburger JJ, Banerjee RK (2011) Toxicity of a biodegradable microneedle implant loaded with methotrexate as a sustained release device in normal rabbit eye: a pilot study. J Ocul Pharmacol Ther 27:151–156. https://doi.org/10.1089/jop.2010.0037

255. Neves SC, Moreira Teixeira LS, Moroni L, Reis RL, Van Blitterswijk CA, Alves NM, Karperien M, Mano JF (2011) Chitosan/poly(e-caprolactone) blend scaffolds for cartilage repair. Biomaterials 32:1068–1079. https://doi.org/10.1016/j.biomaterials.2010.09.073

256. Enea D, Henson F, Kew S, Wardale J, Getgood A, Brooks R, Rushton N (2011) Extruded collagen fibres for tissue engineering applications: effect of crosslinking method on mechanical and biological properties. J Mater Sci Mater Med 22:1569–1578. https://doi.org/10.100 7/s10856-011-4336-1

257. Leonor IB, Rodrigues MT, Gomes ME, Reis RL (2010) In situ functionalization of wet-spun fibre meshes for bone tissue engineering. J Tissue Eng Regen Med 4:524–531. https://doi.org/10.1002/term

258. Nie HL, Ma ZH, Fan ZX, Branford-White CJ, Ning X, Zhu LM, Han J (2009) Polyacrylonitrile fibers efficiently loaded with tamoxifen citrate using wet-spinning from co-dissolving solution. Int J Pharm 373:4–9. https://doi.org/10.1016/j.ijpharm.2009.03.022

259. Meier C, Welland ME (2011) Wet-spinning of amyloid protein nanofibers into multifunctional high-performance biofibers. Biomacromolecules 12:3453–3459. https://doi.org/10.1021/bm2005752

260. De Moraes MA, Beppu MM (2013) Biocomposite membranes of sodium alginate and silk fibroin fibers for biomedical applications. J Appl Polym Sci 130:3451–3457. https://doi.org/10.1002/app.39598

261. Li J, Liu D, Hu C, Sun F, Gustave W, Tian H, Yang S (2016) Flexible fibers wet-spun from formic acid modified chitosan. Carbohyd Polym 136:1137–1143. https://doi.org/10.1016/j.carbpol.2015.10.022

262. Yu DG, Shen XX, Zheng Y, Ma ZH, Zhu LM, Branford-White C (2008) Wet-spinning medicated PAN/PCL fibers for drug sustained release. In: 2nd international conference on bioinformatics and biomedical engineering, iCBBE, pp 1375–1378

263. Majima T, Funakosi T, Iwasaki N, Yamane S-TT, Harada K, Nonaka S, Minami A, Nishimura S-II (2005) Alginate and chitosan polyion complex hybrid fibers for scaffolds in ligament and tendon tissue engineering. J Orthop Sci 10:302–307. https://doi.org/10.1007/s00776-005-0891-y

264. Zhang D (2014) Advances in filament yarn spinning of textiles and polymers. Woodhead Publishing

265. Wieden H, Romatowski J, Moosmueller F, Lenz H (1969) Polyurethane spinning solutions containing ethylene diamine and bis-(4-aminophenyl)-alkane polyurethanes. 6–11

266. Hooshmand S, Aitomäki Y, Norberg N, Mathew AP, Oksman K (2015) Dry-spun single-filament fibers comprising solely cellulose nanofibers from bioresidue. ACS Appl Mater Interfaces 7:13022–13028

267. Zhang C, Zhang Y, Shao H, Hu X (2016) Hybrid silk fibers dry-spun from regenerated silk fibroin/graphene oxide aqueous solutions. ACS Appl Mater Interfaces 8:3349–3358. https://doi.org/10.1021/acsami.5b11245

268. Chang J, Lee Y-H, Wu M, Yang M-C, Chien C (2012) Preparation of electrospun alginate fibers with chitosan sheath. Carbohydr Polym 87:2357–2361. https://doi.org/10.1016/j.carbpol.2011.10.054

269. Han D, Boyce ST, Steckl AJ (2008) Versatile core-Sheath biofibers using coaxial electro-spinning. Mater Res Soc Symp Proc 1094:33–38

270. Lu X, Zhao Q, Liu X, Wang D, Zhang W, Wang C, Wei Y (2006) Preparation and characterization of polypyrrole/TiO_2 coaxial nanocables. Macromol Rapid Commun 27:430–434. https://doi.org/10.1002/marc.200500810

271. Yarin A (2011) Coaxial electrospinning and emulsion electrospinning of core—shell fibers. Polym Adv Technol 22:310–317. https://doi.org/10.1002/pat.1781

272. Yu D, Branford-White K, Chatterton N, Zhu L, Huang L, Wang B (2011) A modified coaxial electrospinning for preparing fibers from a high concentration polymer solution. Express Polym Lett 5:732–741. https://doi.org/10.3144/expresspolymlett.2011.71

273. Granero A, Razal J, Wallace G, Panhuis M (2010) Conducting gel-fibres based on carrageenan, chitosan and carbon nanotubes.pdf. J Mater Chem 20:7953–7956. https://doi.org/10.1039/c0jm00985g

274. Nohemi R, Araiza R, Rochas C, David L, Domard A (2008) Interrupted wet-spinning process for chitosan hollow fiber elaboration. Macromol Symp 266:1–5. https://doi.org/10.1002/masy.200850601

275. Kou L, Huang T, Zheng B, Han Y, Zhao X, Gopalsamy K, Sun H, Gao C (2014) Coaxial wet-spun yarn supercapacitors for high-energy density and safe wearable electronics. Nat Commun 5:3754. https://doi.org/10.1038/ncomms4754

276. Gupta V (1997) Solution-spinning processes. In: Gupta VB, Kothari VK (eds) Manufactured fibre technology. Chapman & Hall, London, pp 124–138
277. Buragotiain C, Vojta M, Uchtyama Y, Nohara M, Ino T, Terasaki I, Aharony A, Cowtey RA, Yoshiiawa H, Oguchi A, Neto AHC, Imada M, Sandvik W, Neto AHC, Birgeneau RJ, Greven M, Halperin BI, Nelson DR, Kastner MA, Stanley HE, Harris B, Birfieneau RJ, Castro AH (2002) Micro/nano encapsulation via electrified coaxial liquid jets. Science 80(295):1695–1699
278. Reneker DH, Yarin AL (2008) Electrospinning jets and polymer nanofibers. Polym J 49:2387–2425. https://doi.org/10.1016/j.polymer.2008.02.002
279. Wei M, Kang B, Sung C, Mead J (2006) Core-sheath structure in electrospun nanofibers from polymer blends. Macromol Mater Eng 291:1307–1314. https://doi.org/10.1002/mame.200600284
280. Liao IC, Chew SY, Leong KW (2006) Aligned core-shell nanofibers delivering bioactive proteins. Nanomedicine 1:465–471. https://doi.org/10.2217/17435889.1.4.465
281. Moghe AK, Gupta BS (2008) Co-axial Electrospinning for nanofiber structures: preparation and applications. Polym Rev 48:353–377. https://doi.org/10.1080/15583720802022257
282. Su Y, Li X, Wang H, He C, Mo X (2009) Fabrication and characterization of biodegradable nanofibrous mats by mix and coaxial electrospinning. J Mater Sci Mater Med 20:2285–2294. https://doi.org/10.1007/s10856-009-3805-2
283. Fu Y, Kao WJ (2011) Drug Release kinetics and transport mechanisms of nondegradable and degradable polymeric delivery systems. NIH Public Access 7:429–444. https://doi.org/10.1517/17425241003602259.Drug
284. Li Y, Chen F, Nie J, Yang D (2012) Electrospun poly (lactic acid)/chitosan core—shell structure nanofibers from homogeneous solution. Carbohyd Polym 90:1445–1451. https://doi.org/10.1016/j.carbpol.2012.07.013
285. Wongsasulak S, Patapeejumruswong M, Weiss J, Supaphol P, Yoovidhya T (2010) Electrospinning of food-grade nanofibers from cellulose acetate and egg albumen blends. J Food Eng 98:370–376. https://doi.org/10.1016/j.jfoodeng.2010.01.014
286. Zhang Y, Huang Z, Xu X, Lim CT, Ramakrishna S (2004) Preparation of core-shell structured PCL-r-gelatin Bi-component nanofibers by coaxial electrospinning. Chem Mater 12:3406–3409
287. Jalaja K, Naskar D, Kundu SC, James NR (2016) Potential of electrospun core-shell structured gelatin-chitosan nanofibers for biomedical applications. Carbohyd Polym 136:1098–1107. https://doi.org/10.1016/j.carbpol.2015.10.014
288. Xu X, Zhuang X, Chen X, Wang X, Yang L, Jing X (2006) Preparation of core-sheath composite nanofibers by emulsion electrospinning. Macromol Rapid Commun 27:1637–1642. https://doi.org/10.1002/marc.200600384
289. Yang Y, Li X, Cui W, Zhou S, Tan R, Wang C (2008) Structural stability and release profiles of proteins from core-shell poly (DL-lactide) ultrafine fibers prepared by emulsion electrospinning. J Biomed Mater Res 86:374–385. https://doi.org/10.1002/jbm.a.31595
290. Luo C, Stoyanov S, Stride E, Pelan E, Edirisinghe M (2012) Electrospinning versus fibre production methods from specifics to.pdf. Chem Soc Rev 41:4708–4735
291. Li F, Zhao Y, Song Y (2010) Core-shell nanofibers : nano channel and capsule by coaxial electrospinning. In: Kumar A (ed) Nanofibers, pp 419–438
292. McCann JT, Marquez M, Xia Y (2006) Melt coaxial electrospinning: a versatile method for the encapsulation of solid materials and fabrication of phase change nanofibers. Nano Lett 6:2868–2872. https://doi.org/10.1021/nl0620839
293. Cabasso I, Klein E, Smith JK, South G (1976) Polysulfone hollow fibers. I. Spinning and properties. J Appl Polym Sci 20:2377–2394
294. Wienk IM, Scholtenhuis FHAO, Van Den Boomgaard T, Smolders CA (1995) Spinning of hollow fiber ultrafiltration membranes from a polymer blend. J Membr Sci 106:233–243
295. Aptel P, Abidine N, Ivaldi F, Lafaille JP (1985) Polysulfone hollow fibers—effect of spinning conditions on ultrafiltration properties. J Membr Sci 22:199–215. https://doi.org/10.1016/S0376-7388(00)81280-6

296. Britain G, Macmillan MP, Republic GD, Chemistry P, Correns E (1989) Polymer hollow fiber membranes for removal of toxic substances from blood. Prog Polym Sci 14:597–627

297. Polacco G, Cascone MG, Lazzeri L, Ferrara S, Giusti P (2002) Biodegradable hollow fibres containing drug-loaded nanoparticles as controlled release systems. Polym Int 51:1464–1472. https://doi.org/10.1002/pi.1086

298. Lee SH (2011) Microfluidic wet spinning of chitosan-alginate microfibers. Biomicrofluid 5:22208. https://doi.org/10.1063/1.3576903

299. Cited R, City O, Data RUA (2003) Drug releasing biodegradable fiber for delivery of therapeutics. 32

300. Schakenraad JM, Oosterbaan JA, Nieuwenhuis P, Molenam I (1988) Biodegradable hollow fibres for the controlled release of drugs. Biomaterials 9:116–120

301. Greidanus PJ (1990) Biodegradable polymer substrates loaded with active substance suitable for the controlled release of the active substance by means of a membrane. 6

302. Greidanus PJ, Feijen J, Eem'nk MJD, Rieke JC, Olijslager J, Albers JHM (1990) Biodegradable polymer substrates loaded with active substance suitable for the controlled release of the active substance by means of a membrane. US4965128

303. Nelson KD, Crow BB (2003) Drug releasing biodegradable fiber delivery of therapeutics. US7033603B2

304. Ochi R, Nagamine S (2005) Spinneret for wet-spinning acrylic sheath-core compound fiber. WO2005111280A1

305. Stęplewski W, Wawro D, Niekraszewicz A, Ciechańska D (2006) Research into the process of manufacturing alginate-chitosan fibres. Fiber Text East 14:25–31

306. Gupta B, Revagade N, Hilborn J (2007) Poly(lactic acid) fiber: an overview. Prog Polym Sci 32:455–482. https://doi.org/10.1016/j.progpolymsci.2007.01.005

307. Teo WE, Ramakrishna S (2006) A review on electrospinning design and nanofibre assemblies. Nanotechnology 17:R89–R106. https://doi.org/10.1088/0957-4484/17/14/R01

308. Park G, Jung Y, Lee G, Hinestroza JP, Jeong Y (2012) Carbon nanotube/poly (vinyl alcohol) fibers with a sheath-core structure prepared by wet spinning. Fibers Polym 13:874–879. https://doi.org/10.1007/s12221-012-0874-5

309. Li S, Shu K, Zhao C, Wang C, Guo Z, Wallace G, Liu HK (2014) One-step synthesis of graphene/polypyrrole nano fiber composites as cathode material for a biocompatible zinc/polymer battery. ACS Appl Mater Interfaces 6:16679–16686

310. Tsukada S, Nakashima H, Torimitsu K (2012) Conductive polymer combined silk fiber bundle for bioelectrical signal recording. PLoS ONE 7:1–10. https://doi.org/10.1371/journal.pone.0033689

311. Wallace GG, Spinks GM, Leon AP, Teasdale PR (2003) Conductive electroactive intelligent materials systems

312. Cullen DK, Patel AR, Doorish JF, Smith DH, Pfister BJ (2008) Developing a tissue-engineered neural-electrical relay using encapsulated neuronal constructs on conducting polymer fibers. J Neural Eng 5:374–384. https://doi.org/10.1088/1741-2560/5/4/002

313. Li M, Guo Y, Wei Y, Macdiarmid AG, Lelkes PI (2006) Electrospinning polyaniline-contained gelatin nanofibers for tissue engineering applications. Biomater 27:2705–2715. https://doi.org/10.1016/j.biomaterials.2005.11.037

Chapter 2
General Experimental

2.1 Components and Spinning Solutions

2.1.1 Materials

Chitosan [medium molecular weight (MMW)] with a degree of deacetylation of about 80.0%), chitosan [high molecular weight (HMW)] and alginic acid sodium salt from brown algae (medium molecular weight) were obtained from *Sigma-Aldrich Co. LLC*. Acetic acid was supplied from *Ajax Finechem* and used directly without further purification. Calcium chloride fused dihydrate (molecular weight of 147.02 obtained from *Chem-supply*) was used as the coagulating agent. Sodium hydroxide pellets (obtained from *Ajax Finechem*) was also used as the coagulating agent. Isopropanol (ISP) was obtained from *Merck* Chemicals. PEDOT:PSS pellets were purchased from *Agfa* (Orgacon dry™, Lot A6,000 AC) and used as supplied. GO in its Liquid Crystal form was synthesized in house as described in previous studies based on modified Hummers method [1]. L-Ascorbic acid was purchased from *Sigma-Aldrich Co.* Pyrrole was supplied in house and used as obtained.

2.1.2 Gel Spinning Precursors

Chitosan powder 3% (w v^{-1}) was dissolved in water containing 2.5% (v v^{-1}) acetic acid. The powder was dissolved in water and stirred overnight at a temperature of ~50 °C to form a homogenous solution. For the purpose of coaxial spinning, chitosan solution was prepared with three different concentrations of calcium chloride, 0.5, 1 and 2% (w v^{-1}), as the cross-linking agent for the sodium alginate.

© Springer Nature Switzerland AG 2018
A. Mirabedini, *Developing Novel Spinning Methods to Fabricate Continuous Multifunctional Fibres for Bioapplications*, Springer Theses,
https://doi.org/10.1007/978-3-319-95378-6_2

Alginate spinning solution was produced by dissolving alginate powder in water while stirring overnight at ~50 °C to provide homogeneity for the spinning. The sodium alginate in solution was then precipitated in a bath of aqueous calcium chloride (Ethanol/H$_2$O 1:5).

2.1.3 Graphene Oxide Liquid Crystal Dispersion

Liquid Crystal Graphene Oxide (LC GO) dispersions were prepared in house based on the typical previously reported modified Hummers method [1, 2] in which firstly 1 g of EG and 200 mL of sulphuric acid were mixed and stirred in a three neck flask for 24 h. After that, 5 g of KMnO$_4$ was added to the mixture and stirred at room temperature for 24 h. The mixture was then cooled in an ice bath and 200 mL of deionised water and 50 mL of H$_2$O$_2$ were poured slowly into the mixture resulting in a colour change to light brown followed by stirring for 30 min. The resulting dispersion was washed and centrifuged three times with an HCl solution (9:1 vol water: HCl). Repeated centrifugation-washing steps with deionised water were carried out until a solution pH $\geq$ 6 was achieved. Large GO sheets were re-dispersed in deionised water by gentle shaking without the need for sonication. Liquid crystalline (LC) GO dispersions formed spontaneously above a GO concentration of 0.02 w v^{-1}.

2.1.4 PEDOT:PSS Dispersion

Aqueous dispersions were made from PEDOT:PSS pellets holding the concentrations of up to 2.5% w v^{-1}. The dispersions were then undertaken a homogenising step (Labtek IKA® T25) at 18,000 rpm for 15 min followed by 1 h bath sonication (Branson B5500R-DTH). Another PEDOT:PSS solution was also made using the previously explained procedure. However, Polyethylene glycol (PEG) (Mw 2000) (10% w v^{-1}) was finally added to the formulation directly, homogenised at 18,000 rpm for more 2 min followed by 10 min bath sonication.

Wet-spinning of formulations prepared only by mechanical stirring or homogenization without further sonication would cause relatively bigger particles in the dispersion to be accumulated at the needle tip and formed a frequent breaking during fibre formation (clearly visible in the fibre structure inside the coagulation bath). The formulations were furthered filtered with 1 μm sterile filters.

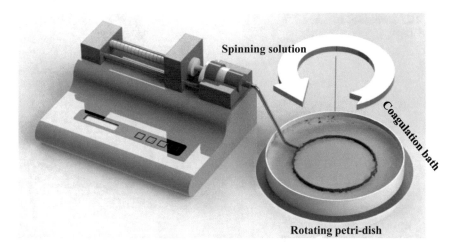

Fig. 2.1 A schematic of rotary wet-spinning method

2.2 Experimental Methods

2.2.1 Spinning Techniques

Two types of fibre wet-spinning methods were utilized, rotary wet-spinning and long bath for coaxial wet-spinning.

2.2.1.1 Rotary Wet-Spinning Approach

Pure hydrogel and GO fibres were spun *via* a rotary wet-spinning method. For this purpose, the spinning solutions were injected into a rotating petri-dish contained an appropriate coagulation bath through the thin nozzle of the spinneret as depicted in Fig. 2.1. In addition, to demonstrate continuous fibre wet-spinning, spinning trials were carried out using a custom-built wet-spinning setup (depicted in Fig. 1.8, previously).

2.2.1.2 Coaxial Wet-Spinning

Core-sheath fibres were successfully spun using a coaxial spinneret. For this purpose, a novel coaxial spinneret with two input ports was designed (schematically shown in Fig. 2.2). Spinning solution of the core component was injected through port B and extruded through the centre outlet nozzle into the proper coagulation bath. Simultaneously, the sheath spinning solution was extruded as the sheath of the fibre,

Fig. 2.2 A schematic of
coaxial spinneret

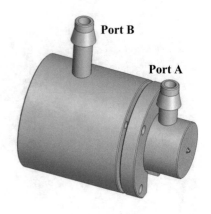

Port B

Port A

providing an outer casing for the core, by injection through port A which facilitates
extrusion through the outer segment of the spinneret nozzle.

A schematic of the coaxial wet-spinning setup was previously shown in Fig. 1.11.
The setup consists of two injection syringes and pumps (*KDS100, KD Scientific Inc.*),
connected to port A and B of the predesigned coaxial spinneret, a coagulation bath
and a stretching collector (*Nakamura Service Co.*). Also, the rates at which spinning
solutions are injected with are quite important in the successful formation of coaxial
fibres. Generally speaking, the core component is needed to be delivered at a lower
rate compared to the sheath material because it is needed to have sufficient time to
be covered by the sheath material.

2.2.1.2.1 Calculation of Output Flow Rate

Considering V_i for each component the outlet sectional area diameter, the output
flow rate was calculated according to the equation, $Q = VA$ [3] where Q represents
the volumetric flow rate (m^3/hr), V the linear velocity (mL/hr) and A the cross-
sectional area of spinneret head (m^2) during coaxial spinning.

Assuming a Newtonian flow, the shear rate during injection can be estimated from
the flow rate (Q) and the inner radius of the spinneret (R), using Eq. 2.1 as below:

$$\gamma = \frac{4Q}{\pi R^3} \tag{2.1}$$

Consequently, the viscosities of core and sheath solutions during the process could
be approximately calculated.

2.2.1.3 Rheological Characterisation of Spinning Solutions

An understanding of the rheological properties of spinning solutions is essential to determine the optimum conditions for wet-spinning. This feature is regarded as the primary criterion for the selection of suitable concentrations of materials for the purpose of coaxial spinning. Viscosity changes in spinning solutions were measured in flow mode (cone and plate method) (angle: 2°, diameter: 60 mm) by Rheometer-(AR-G2 TA Instruments, USA) and repeated three times for each sample. Approximately 2 mL of spinning solutions/dispersions was loaded into the rheometer plate carefully not to shear or stretch the sample. Shear viscosity measurements were carried out three times for each sample at room temperature (~25 °C).

2.2.1.4 Stereomicroscope Observation

Digital images and measurements of prepared fibres (in their wet/dry states) were obtained. The fibre diameters were also measured using a LEICA M205 stereomicroscope and LAS 4.4 software.

2.2.1.5 Low Vacuum Scanning Electron Microscopy

The morphology of fibres, surface and cross-sectional structure, were examined in a JSM-6490LV SEM. Samples were prepared for imaging by immersion in SBF (for about 30 min) beforehand and then short lengths (about 5 mm) were removed, drained of excess medium and inserted into holes (1 or 1.5 mm diameter) which had been pre-drilled into a small brass block (approximately 25 mm diameter × 10 mm) (See Fig. 2.3a). The holes allow the fibres to be inserted upright and protrude from the brass block (See Fig. 2.3b). The block containing the mounted fibres was then plunged into liquid nitrogen for about 45 s and a liquid nitrogen cooled razor blade was run across the surface of the block to fracture the fibres. The block was then quickly transferred to the LVSEM for examination. SEM images were taken in HV mode at 15 kV operating voltage and a spot size setting of 45. Due to their inherent water content, the fibres remained conductive for a period of 25–30 min in the SEM vacuum and so no coating was required.

2.2.1.6 Scanning Electron Microscopy

The morphological structures of the fibre surface and cross-sections were observed by a JEOL JSM-7500 FESEM. Samples for imaging were prepared by cutting cross sections in liquid nitrogen using a scalpel blade. They were then coated (EDWARDS Auto 306) with a thin (10 nm) layer of Pt to aid with imaging and minimise beam heating effects. Cross-sections were analysed at 25 kV accelerating voltage and a spot size setting of 13 under High Vacuum (ultimately is of the order of 0.1 mPa).

(a)   **(b)**

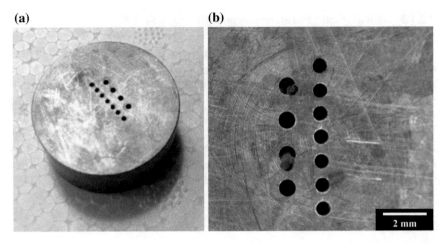

Fig. 2.3 a 25 mm × 10 mm pre-drilled brass block, **b** wet fibres inserted upright into the holes and protrude from the brass block to be imaged under LVSEM

2.2.1.7 Fourier Transform Infrared

The Fourier transform infrared (FTIR) spectroscopy was employed to examine which functional groups might be present in the structure. FTIR spectroscopy was performed in KBr pellet on a Shimadzu FTIR Prestige-21 spectrometer, in the 700–4000 cm^{-1} range with 4 cm^{-1} resolution.

2.2.1.8 Dynamic Mechanical Analysis

The mechanical properties of fibres were measured *via* performing tensile experiments. Tensile tests were carried out on a Dynamic Mechanical Tester (EZ-L Tester from Shimadzu, Japan), at 2 mm min^{-1} and a gauge length of 10 mm. Average values of tensile strength and maximum strain were determined after repeating the test three times.

2.2.1.9 Thermogravimetric Analysis

Thermogravimetric Analysis (TGA) was used to determine the decomposition temperature of both materials in produced fibres. As a result, the weight ratio of each hydrogel involved in forming the fibres can be roughly determined. Thermogravimetric measurements were made using a TA instruments Thermogravimetric Analyser (TGA) model TGA/SDTA851e. The temperature range studied was 25–800 °C at the heating rate of 2 °C min^{-1} under an air atmosphere. The mass of the sample pan was continuously recorded as a function of temperature.

2.2.1.10 UV-Visible Spectroscopy

An absorbance calibration curve for sample analysis was constructed using a Shimadzu UV 1601 spectrophotometer. UV–vis spectra of SBF solutions containing TB with different concentrations were recorded between 200 nm and 1100 nm. The details of the experiments are given in Sect. 3.2.3.2.

2.2.1.11 Cyclic Voltammetry

Cyclic voltammetry was used to investigate the redox/electrochemical properties and determine the specific capacitance of as-prepared fibres. The experiment was performed on different fibre types inside both an aqueous and organic solutions of 1 Molar aqueous NaCl and phosphate buffer saline (PBS) and 0.1 M tetrabutylammonium bromide (TBABF$_4$), respectively. A three-electrode cell composed of the coaxial fibre as the working electrode, an Ag/AgCl reference electrode (for aqueous electrolytes) and a Pt mesh as an auxiliary electrode was utilized and an E-Corder 401 interface and a potentiostat (EDAQ) was employed. In case of organic electrolytes, 0.1 M TBABF$_4$ in acetonitrile was used as the reference electrode. Steel wire with the average diameter of 25 μm was inserted into the fibre core while spinning. Each data point in the EC plots is an average result of three experiments.

In order to investigate the electrical storage and retention characteristic of the coaxial fibres, the specific capacitance can be calculated from capacitance versus time (C–t) curves achieved from the CV curve using the Eq. 2.2 given below [4]:

$$C = \frac{Q}{2 \times \Delta V \times m} \qquad (2.2)$$

where Q is the average integral area of CV curve, ΔV is the working voltage window and m is mass (g) of the electrode material which get gravimetric capacitance in F g^{-1}.

2.2.1.11.1 Insertion of Cotton-Steel Wire into Coaxial Fibre

To make a connection to the conductive core of the coaxial fibres, a cotton-steel wire with the average diameter of 20 μm was inserted into the fibre core during the spinning process. The flexible cotton-steel thread assembled into the core showed very low resistivity with an insignificant impact on the CV results (as shown is CV curves later on) providing also a conductive pathway for electrochemical measurement. To this end, a novel triaxial spinneret with three input ports was designed (schematically shown in Fig. 2.4. The cotton-steel fibre was inserted into the port C which is embedded at the back of spinneret while spinning solution of the core was injected through port B and the sheath component injected through port A, simultaneously.

(a)   (b)   Port B

Port C   Port A

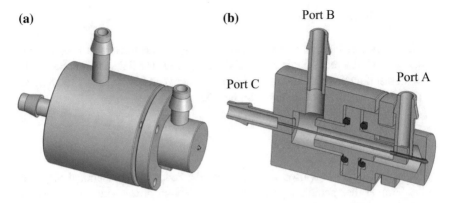

Fig. 2.4 Schematic of **a** lateral view and **b** cross-section of the triaxial spinneret

2.2.1.12   Conductivity Measurements

To determine electrical conductivity, a potential difference is applied across a sample material and the current flow measured based on "Ohm's law" which is indicated in Eq. 2.3.

$$I = V/R \tag{2.3}$$

Then, the conductivity of a material (σ) is defined as the reciprocal of the resistivity, according to the Eq. 2.4;

$$\sigma = \frac{1}{\rho} = \frac{1}{R_s d} \tag{2.4}$$

where ρ is the resistivity of the sample and d is the diameter of the fibre. To calculate the number, a linear in-house built four-point probe conductivity set-up with equal probe spacing of about 2.3 mm was employed to measure the room temperature conductivity of the monofilament fibers using a galvanostat current source (Princeton Applied Research Model 363) and a digital multimeter (HP Agilent 34401A). It is necessary for the fibres to entirely attach to the probes using silver paint. Conductivity is calculated from the surface of the fibres. Therefore, this method did not seem to be applicable on coaxial fibres.

2.2.1.13   In Vitro Bioactivity Experiments

The cytotoxicity, biocompatibility and proliferation of cells on the fibres were determined for different types of cells. Cell adhesion and proliferation on as-prepared fibres were evaluated without adding extra-cellular matrices to the fibres. Prior to use fibres in cell culture, about 30 mm lengths of fibres were fixed onto microscope slides with 4-well chamber wells (Lab Tek®II, Thermo Fisher Scientific) glued on top which were then allowed to get dry overnight. Then, the samples underwent a sterilization process consisted of two washes in a sterile condition of 70% ethanol (each for 30 min) and then four washes (each for 30 min) in sterile phosphate buffered saline (PBS, *Sigma-Aldrich Co.*). Finally, they were kept in cell culture media overnight to remove all excess acid.

References

1. Jalili R, Aboutalebi H, Esrafilzadeh D, Shepherd R, Chen J, Aminorroaya-yamini S, Konstantinov K, Minett A, Razal J, Wallace G (2013) Scalable one-step wet-spinning of graphene fibers and yarns from liquid crystalline dispersions of graphene oxide: towards multifunctional textiles. Adv Funct Mater 23:5345–5354. https://doi.org/10.1002/adfm.201300765
2. Aboutalebi SH, Gudarzi MM, Bin Zheng Q (2011) Spontaneous formation of liquid crystals in ultralarge graphene oxide dispersions. Adv Funct Mater 21:2978–2988. https://doi.org/10.1002/adfm.201100448
3. Subramanian RS (2012) Pipe flow calculations
4. Yu D, Goh K, Wang H, Wei L, Jiang W, Zhang Q, Dai L, Chen Y (2014) Scalable synthesis of hierarchically structured carbon nanotube–graphene fibres for capacitive energy storage. Nat Nanotechnol 9:555–562. https://doi.org/10.1038/NNANO.2014.93

Chapter 3
Preparation and Characterisation of Novel Hybrid Hydrogel Fibres

3.1 Introduction

Tissue scaffolds can be synthesised from synthetic or biologically based bioresorbable polymers that act as a functional or inert framework for missing or malfunctioning human tissues and organs. The primary role of a scaffold is to provide a temporary substrate to which cells can adhere [1]. A scaffold should ideally promote attachment, migration, proliferation and differentiation of cells, as appropriate [1]. To achieve this, scaffolds must have certain physical characteristics tailored to tissue requirements, such as porosity, structural integrity, defined pore size and controlled degradability. Furthermore, the scaffold should provide the optimal biochemical microenvironment for the seeded cells.

Biopolymers are commonly used in clinical and biological applications for tissue engineering purposes [2–16]. Chitosan and alginate are promising, naturally occurring, biopolymers. They possess several unique properties including biodegradability, biocompatibility, low toxicity, low immunogenicity and antimicrobial activity [17]. These biopolymers can be obtained from different sources such as microbial and animal [18] and are used in various applications such as wound dressings [19–23], tissue scaffolds [24–31], cell encapsulation [32, 33], drug delivery [34–40] and also as cell delivery vehicles [41]. Due to their simple processability, they can be fabricated in various forms such as gels, fibres, nano and microspheres and scaffolds [5, 42]. An ongoing quest of several investigators is focused towards tissue engineering and repair through the transplantation of cells seeded onto biodegradable polymer scaffolds [43–45]. Porous chitosan matrix has been used as a scaffold for skin [6, 9, 46], liver [14], bone and cartilage [2–4, 7] cardiac [10, 15, 47], corneal [48] and

Parts of this chapter have been reproduced with permission from https://doi.org/10.1002/mame.201 500152 (A. Mirabedini, J. Foroughi, T. Romeo, G. G. Wallace, "Development and characterisation of novel hybrid hydrogel fibres" Macromolecular Materials and Engineering, Volume 300 (12), pages 1217–1225, 2015).

vascular regenerative tissue remodelling [5, 49, 50]. Alginate has been extensively used as a scaffold for liver [11], bone [51], nerve [52] and cartilage engineering [53].

Chitosan and alginate are two of the most important members of the "polysaccharides from non-human origin" group [54]. Chitosan is a cationic polymer derived from crustacean skeletons, while alginic acid is an anionic polymer, typically derived from brown algae [55]. Alginate is a linear, binary copolymer composed of 1, 4-linked β-D-mannuronic acid (M) and α-L-guluronic acid (G) monomers. The composition of alginate (the ratio of the two uronic acids and their sequential arrangements) varies with the source. Salts of alginic acid with monovalent cations such as sodium alginate are all soluble in water [56]. The high acid content allows alginic acid to undergo spontaneous and mild gelling in the presence of divalent cations, such as calcium ions. However, calcium alginate fibres have proven to be unstable structures as tissue scaffolds or drug vehicles for in vivo usages due to its tendency to swell and dissolve in various ionic solutions which are present in the body [57, 58]. Structurally, chitosan is a semi-synthetically derived aminopolysaccharide which is the N-deacetylated product of chitin, i.e. poly-$(1 \rightarrow 4)$-2-amino-2-deoxy-β-D-glucose [8, 55, 59, 60]. This hydrogel is highly reactive due to free amine groups and is readily soluble in weakly acidic solutions resulting in the formation of a cationic polymer with a high charge density [61–64]. Chitosan can be produced in a variety of forms including films, fibres, nanoparticles and microspheres. Nevertheless, the strong alkaline condition (pH > 12) needed to form chitosan-based structures, can limit its utilization for loading most of the drugs or bioactive molecules into it. Chitosan can also form ionic complexes with water-soluble anionic polymers such as alginate. Therefore, there has been an increased interest in fabricating alginate-chitosan polyion systems for several bioapplications such as cartilage and bone regeneration [65]. Both alginate and chitosan have been approved by the FDA for clinical use. Incorporating these two materials together in the one structure allows us to form polyanion–polycation complexes [66]. The alginate/chitosan polyelectrolyte complex has combined properties of the two individual components, such as more stable to pH change for shape-keeping than alginate or chitosan alone in aqueous medium [67] alongside providing the opportunity to use it as drug-loaded bioscaffolds. Chitosan-alginate hybrid polymer structures have also been shown to promote biological responses including enhancing cell attachment and proliferation [68].

Synthetic fibres are usually made *via* one of four typical spinning methods - wet-spinning, melt-spinning, dry-spinning or electrospinning. Wet-spinning is the oldest among the four processes and is generally used to produce natural fibres that cannot be formed by either melt or dry-spinning [59]. Wet-spinning is the preferred approach for both chitosan [69, 70] and alginate [71] fibres because it maintains the strong interchain hydroxyl forces in their chemical structure.

Wet-spun fibres have been widely used for controlled release applications. Dyes are often used to investigate the loading and release properties of polymeric matrices due to the structural resemblance with many low molecular drugs [72–75]. Ding et al. showed that the loading and release of methylene blue (MB) from polyelectrolyte multilayers constructed using poly (diallyl dimethylammonium chloride) and poly(acrylic acid) (PAA) strongly depended on the pH and ionic strength of the external media [76]. Graciano et al. also evaluated the potential of chitosan gels delivery systems containing Toluidine blue O (TBO or TB) for use in the photodynamic therapy of buccal cancer treatment [77]. TB is a chromophore in the phenothiazinium family with a strong absorption band in the 620–660 nm region, which is located within the phototherapeutic window (600–750 nm) in which light penetration into the tissue is maximized [77]. TB was used as an indicative dye incorporated into the core component of coaxial fibres to study the release kinetics of the fibres.

In this chapter, a new wet-spinning approach is developed that employs a coaxial spinneret for production of chitosan/alginate (Chit/Alg) core-sheath fibres. This research explores the conditions necessary to achieve optimal properties of the chitosan-alginate core-sheath fibre in one-step spinning. The secondary aim of this research was to investigate the mechanical and swelling properties of these fibres in order to optimize them for use in biomedical applications. Finally, the release kinetics of coaxial fibres have been studied and compared with its forming constituents. Significantly improved mechanical and swelling properties of coaxial alginate-chitosan fibres, compared to their alginate counterparts were reported.

3.2 Experimental

3.2.1 Materials

Medium molecular weight chitosan was used in this study as the core spinning solution. Toluidine Blue O (also known as toluidine blue or TB) as an indicator ingredient was supplied from *Sigma-Aldrich Co.* for the release experiments.

A simulated body fluid (SBF) solution, with ion concentrations approximately equal to those of human blood plasma, was prepared with the following composition; 142 mM Na^+, 5 mM K^+, 1.5 mM Mg^{2+}, 2.5 mM Ca^{2+}, 10^3 mM Cl^-, 27 mM HCO_3^-, 1.0 mM HPO_4^{2-} and 0.5 mM SO_4^{2-} with the final adjusted pH of 7.4 ± 0.05 [26]. This solution was used as an aqueous medium to re-swell the dried fibres for imaging.

3.2.2 Wet-Spinning of Chitosan, Alginate and Chit-Alg Coaxial Fibres

The spinning solutions were prepared as described in Sect. 2.1.2. Wet-spun chitosan fibres were produced in a coagulation bath consisting of 1 M sodium hydroxide (NaOH) (Ethanol/H_2O 1:5) using a rotary wet-spinning system. Uniform alginate fibres were spun in a 2% $CaCl_2$ coagulation bath. Core-sheath fibres of chitosan-alginate (Chit-Alg) were successfully spun using a coaxial spinneret. For this purpose, a novel coaxial spinneret described was used. The chitosan spinning solution (with different amounts of $CaCl_2$) was injected as the core component and extruded through the centre nozzle into the coagulation bath of calcium chloride (Ethanol/H_2O 1:5). Simultaneously, alginate was injected as the sheath of the fibre, providing an outer casing for the core, by injection through port A. In this method, by using a blend of chitosan with various percentages of calcium chloride, it is possible that the alginate sheath can be coagulated from the inner chitosan core, while also creating the opportunity to react chitosan with sodium alginate at a much faster rate. The setup is shown in Fig. 1.11, previously. Therefore, as mentioned earlier chitosan solutions including 0.5, 1 and 2% (w v^{-1}) $CaCl_2$ were prepared for the core component of the fibres and alginate the sheath. The samples are named here as Chit-Alg (0.5), Chit-Alg (1) and Chit-Alg (2). Solutions were delivered at flow rates of 14 mL h^{-1} for chitosan and 25 mL h^{-1} for the sheath. Coaxial fibres of Chit-Alg (1) were then used for all imaging characterisation tests.

TB was used as an indicative dye incorporated into the coaxial fibres to track the release experiment. For the purpose of fibre preparation for release experiments, the dye was mixed with chitosan solution before spinning with the concentration of 0.1% (w v^{-1}) and then injected as the core component. These solutions were then spun into the same coagulation baths which were previously used to make pristine fibres. Coaxial fibres containing TB were also fabricated using the method mentioned previously with the small difference of using chitosan/TB solution as the core component.

3.2.3 Characterisation Methods

3.2.3.1 Cell Viability Experiments

All tissue collection and handling in this chapter of my thesis was performed according to St. Vincent's Hospital approved protocols by Dr. Anita Quigley. Myoblasts were prepared from human or mouse (C57Bl6) muscle tissue as described previously [70]. Purified myoblasts were expanded and maintained in proliferation media (HamsF10, Trace Biosciences) supplemented with 20% fetal bovine serum (Invitrogen), 2.5 µg mL^{-1} bFGF (PeproTech), 2 mM L-glutamine, 100 U mL^{-1} penicillin and 100 µg mL^{-1} streptomycin (Invitrogen). All cell culture was carried out in 5% CO_2 at 37 °C. Fibres were briefly sterilized in 70% ethanol before rehydration in MilliQ water. Fibres were then equilibrated in HamsF10 and proliferation media in 12 well plates before human or rodent myoblasts were seeded at a density of 30,000 cells mL^{-1} onto the fibres. Myoblasts were allowed to attach for 3 days before analysis for cell attachment and survival.

Cells were stained with 1 µM Calcein AM (Life Technologies) and 0.5 µM propidium iodide for 15 min at 37 °C (*Sigma-Aldrich Co.*) to indicate viable and dead cells. Digital images were obtained under fluorescence using an IX70 inverted microscope (Olympus) and Spot (version 4.7.0) software (Diagnostic Instruments).

3.2.3.2 TB Release Measurement

The release kinetics of the prepared fibres for drug release applications was studied using TB as a model dye introduced into the fibres over a 5-day period. The amount of released TB was determined *via* UV–vis spectroscopy by monitoring the absorption of TB at its λ_{max} 630 nm in SBF (pH 7.4, total volume is 1 mL). To construct an absorbance calibration curve for sample analysis using a Shimadzu UV 1601 spectrophotometer, UV–vis spectra of SBF solutions containing TB with different concentrations were recorded between 200 nm and 1100 nm. Approximately 5 cm of each dried sample (in triplicate) was placed in a 2 mL Eppendorf tube and 1 mL of SBF was added into it. The release medium was taken by micro-pipette at specific time points over 5 days and replaced with the same volume of fresh SBF solution to maintain the total volume constant. The percentage of released TB (%) was plotted versus time.

3.3 Results and Discussion

Initial investigations were aimed at determining the spinnability and the physical characteristics of potential dopes to ensure they were in the range required for fibre formation.

3.3.1 Spinnability Versus Concentration

Spinnability can be defined as the ability of a material of being suitable for spinning or the capability of being spun. In the context of wet-spinning, spinnability could be referred to the ability of a solution to form fibrillar arrangements *via* injection into a non-solvent medium which makes it precipitate, so-called a coagulation bath [78]. The spinnability of a polymer solution depends on many parameters, including the rheological properties of a solution, size of nozzle, shear rate applied during injection through spinneret and mass transfer rate difference between the extruded solution and non-solvent. Several types of either suitable solvent or coagulant could be employed depending on the chemical structure of the material. Often an upper and lower limit for polymer concentration during wet-spinning is considered facilitates spinning continuous length of fibres. The capillary break-up is widely understood as a surface-tension induced break-up of filaments into drops can occur at low concentrations of the polymer solution which determines the lower limit of spinnability [79]. Spinnable concentrations have been reported for chitosan varying from 2–15% (w v^{-1}) [17–19, 22, 23]. Here, we found that 2–5% (w v^{-1}) is the appropriate concentration range enabling wet-spinning of MMW chitosan into a coagulation bath of 1 M NaOH. Observations also indicated that aqueous alginate solutions at a concentration of below 2% (w v^{-1}) would not generate a continuous fibrous structure *via* wet-spinning;

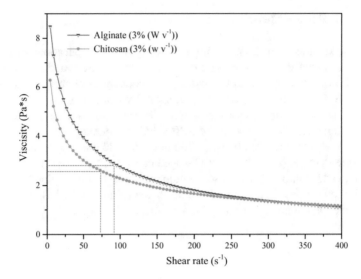

Fig. 3.1 Viscosities of spinning solutions of chitosan and sodium alginate

increasing the alginate concentration from 2 to 4% (w v^{-1}), the solution became highly spinnable. Then again, at concentrations above 4% (w v^{-1}), the solution became highly viscous which impede continuous flow through the needle, rendering the solution unspinnable. A concentration of 3% (w v^{-1}) has been thus selected for both gel precursors due to the ease of spinnability, together with maintaining the suitable mechanical properties for coaxial wet-spinning. In brief, the concentrations of spinning solutions were kept constant to study the effect of spinning parameters such as injection rates and the amount of calcium chloride inside the coagulation bath as well as amount mixed with chitosan solutions.

3.3.2 Rheology

Viscosity is considered in the selection of suitable concentrations of chitosan and alginate solutions for fibre spinning. For coaxial spinning matching viscosities of the two components is also a consideration [68].

Figure 3.1 shows changes in viscosity versus shear rate was determined from aqueous solutions of chitosan at 3% (w v^{-1}) and alginate at 3% (w v^{-1}). Spinning solutions of 3% (w v^{-1}) chitosan resulted in a solution with a viscosity of 6.4 Pa*s. The viscosity of 3% (w v^{-1}) sodium alginate solution was approximately 8.5 Pa*s.

The viscosities of the two solutions became closer as the shear rate increased. Under shear, hydrogel chains are in a less expanded conformation and become less entangled causing the viscosity to drop. At the time of spinning, chitosan is injected with the rate of 14 mL h^{-1} while Vi is 25 mL h^{-1} for the alginate solution. Considering

Fig. 3.2 The capability of producing an unlimited length of coaxial Chit-Alg (1) fibres as shown onto a collector

the outlet sectional area diameter, the output shear rate (γ) was calculated according to calculations in Chap. 2, Sect. 2.2.1.2.1. The shear rates calculated to be about ~97 S^{-1} for alginate and ~75 S^{-1} for chitosan solutions which resulted in a viscosity of ~2.5 Pa*s for chitosan and ~2.8 Pa*s for alginate solutions. These outcomes seem to be ideal for coaxial spinning.

3.3.3 Continuous Spinning of Coaxial Fibres

To produce continuous uniform fibres, chitosan with the injection rate of 14 mL h^{-1} and alginate solution with the rate of 25 mL h^{-1} have been simultaneously injected into the 2% (w v^{-1}) aqueous $CaCl_2$ coagulation bath through the ports built in the coaxial spinneret. The spinning method to achieve the optimal properties was described in Sect. 2.2.1.2 in detail, previously. Using this method, unlimited length of fibres could be obtained which was collected using a collector as shown in Fig. 3.2. It is worth mention that the preparation of coaxial fibres without incorporating a certain amount of $CaCl_2$ did not turn out to be successful as tried.

The woven structure of Chit-Alg fibres (containing TB) in a dry and wet state were shown in Fig. 3.3a and b, respectively. This capability provides the potential for these structures to be utilized as tissue scaffolds and drug delivery vehicles applications.

3.3.4 Morphology of As-Prepared Fibres

The stereomicroscope images of wet-spun chitosan, alginate and core-sheath Chit/Alg fibres are shown in Fig. 3.4 in wet and dry-states.

(a)   **(b)**

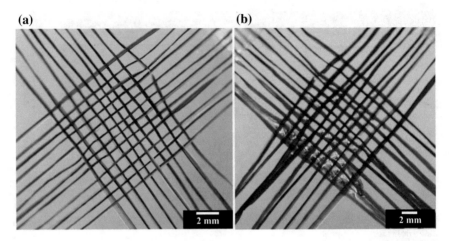

Fig. 3.3 The photographs of scaffold structure woven by coaxial fibres; imaged in **a** dry and **b** wet-state

As can be seen in Fig. 3.4a, b, the surface of chitosan fibre seemed to be very smooth and soft, while some wrinkles can be noticed spreading on the surface of the alginate fibre which will be increased during the fibre drying process. Moreover, one can see that the core-sheath fibres are straight and smooth with a core loaded in the centre of fibres, have a uniform structure and diameter of ca. 220 μm and 136 μm for the sheath and core when wet, respectively (Fig. 3.4c). However, the dried fibres hold the thickness of ~140 ± 10 μm (This average value was calculated after measuring the diameter under the stereomicroscope ten times). In addition, some lines or longitudinal indentations can be observed running parallel with the fibres on totally dried core-sheath fibre as shown in Fig. 4.3d. The chitosan core is ~90 μm which is surrounded by a thin layer of alginate sheath of ~8–12 μm. Still, the thicknesses of core and sheath materials are adjustable by changing solution feed rates and the drawing ratio (data, variables vs. dimensions). Changes in dimensions of core and sheath components for three various injection rates and two drawing ratios as a result of changing the collection rates) are presented in Table 3.1 (determined from LV-SEM images). Considering two selected collection rates at angular velocities of 20 and 60 rpm (with the assumption of keeping the injection rates constant), it is possible to measure draw ratio upon increasing the collection rate from 20 to 60 rpm while using the same collector (constant radius). The relationship between linear velocity and angular velocity could be defined with the Eq. 3.1;

$$v = \omega r \qquad (3.1)$$

where ω, is the angular velocity (collection rate), v is the linear velocity (draw ratio) and r is the radius of the collector.

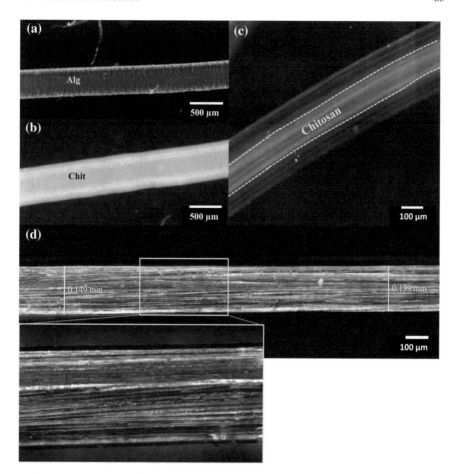

Fig. 3.4 Stereomicroscope images of the side view of wet **a** alginate, **b** chitosan, **c** coaxial Chit-Alg (1) and **d** dry Chit-Alg (1) fibre

Table 3.1 Thickness of sheath and core, μm, in wet-state as a function of wet-spinning condition

Sheath/Core injection rates (mL h^{-1})	Draw Ratio	
	1	3
25/14	220 ± 6.1, 136 ± 2.05	152 ± 4.3, 100 ± 7.5
40/14	231 ± 3.2, 103 ± 8	165 ± 1.2, 107 ± 7.6
25/20	Not successful	

As could be seen in Table 3.1, the fibre diameter decreased as the drawing ratio increased. In general, the molecular orientation of fibre materials obtained through the drawing process governs their properties, particularly the mechanical properties. In addition, the thickness of the sheath would have a direct relationship with increasing

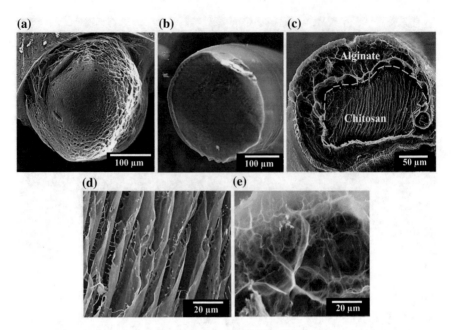

Fig. 3.5 LV-SEM images of hydrated as-prepared **a** alginate, **b** chitosan and **c** Chit-Alg (1) cross section in SBF, **d** chitosan core arrangement in cross section **e** alginate sheath construction in the cross section

its injection rate; the thickness of alginate increased as the shear rate increased while the feeding rate of core component was kept constant. However, while the application of shear is essential in obtaining orientation in the fibre, high shear rates develop beaded non-uniform fibre in the coagulation bath as a result of die swell (swelling of the free jet of solution upon injection from spinneret) and skin formation. Die swell occurs as a consequence of polymer relaxation due to its low entropy conformation after shear is applied during extrusion through the spinneret, where polymer molecules are oriented by the flow. The diameter of the jet then decreases as a result of drawing along the spinning path. A hard skin is also formed on the surface of the filament which results in the rate the jet diameter decreases. When the shear rate of chitosan increased to 20 mL h^{-1}, the formation of the coaxial structure did not turn out to be successful. It seemed that the sheath components were not thick enough to hold the core material in place.

LV-SEM images of cross-sections and the surfaces of solid and core-sheath fibres are illustrated in Fig. 3.5a–e. They give valuable information about the morphology of the two polymers. Before imaging fibres were immersed in SBF and imaged with LV-SEM in an attempt to capture structural information in the "wet-state", since that is how they would be used in future possible applications.

Cross-sections of solid fibres fabricated based on spinning conditions mentioned in Sect. 3.2.2, clearly show the cylinder shaped form of the hydrated chitosan and

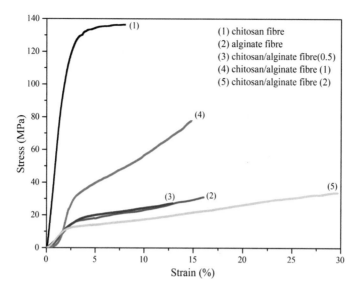

Fig. 3.6 Stress–strain curves obtained from tensile tests of alginate single and chitosan/alginate coaxial fibres using different CaCl₂ concentrations

alginate solid fibres (Fig. 3.5a and b, respectively). Alginate fibres appeared to be permeable and spongy, while the cross-section of chitosan fibres appeared to be denser. In contrast, cross-sections of the coaxial fibres reveal slightly irregular, oval shaped fibres with a distinct separation between chitosan in the core and the outer alginate sheath as is indicated in Fig. 3.5c. In addition, both polymers show an extensive porous structure in the coaxial structure. On the cross-section of chitosan, regular crystalline structures can be seen which are probably due to the presence of calcium chloride inside the core (Fig. 3.5d), while alginate has a honeycomb structure (Fig. 3.5e). It is evident that the fibre is composed of two distinct areas of chitosan and alginate.

3.3.5 Mechanical Properties of As-Prepared Fibres

The mechanical properties of alginate, chitosan and Chit-Alg coaxial fibres employing different concentrations of calcium chloride in chitosan core spinning dope are depicted in Fig. 3.6. Ultimate stresses (MPa), ultimate strains (%), Young's moduli (MPa) and swelling ratios (%) were measured for alginate, chitosan, Chit-Alg (0.5), Chit-Alg (1) and Chit-Alg (2) fibres, respectively.

Mechanical analysis results revealed that with the addition of more calcium chloride to the core-dope, Young's modulus decreased. Increasing the amount of calcium chloride into fibre core will probably cause agglomerations which can lead to phase

Table 3.2 Comparison of mechanical properties of solid and coaxial biofibres

Sample	Breaking stress (MPa)	Strain at break (%)	Young's modulus (GPa)	Initial swelling ratio (%)
Alginate fibre	~31 ± 5	~26 ± 3	1.6 ± 0.15	(Non-measurable)
Chitosan fibre	~146 ± 30	~19 ± 5.2	6.6 ± 0.8	~90
Chit-Alg (0.5)	~30 ± 5	~22 ± 8	0.6 ± 0.1	~360
Chit-Alg (1)	~80 ± 10	~14 ± 3	1.9 ± 0.2	~540
Chit-Alg (2)	~28 ± 5	~37 ± 5	0.55 ± 0.1	~385

separation. Thus, there would be an upper threshold for the amount of $CaCl_2$ in the core at which the optimum mechanical parameters could be achieved. As a result, the mechanical properties of as-prepared fibres such as Young's modulus and ultimate stress were decreased by the addition of more than 1% (w v^{-1}) $CaCl_2$.

The results, which are presented in Table 3.2, also confirmed the reinforcing role played by the chitosan core in coaxial Chit-Alg fibres. Young's modulus was measured to be *ca.* 1.7 and 6.6 MPa for alginate and chitosan solid fibres, respectively. It has been also revealed that the fibres which contain 1% (w v^{-1}) $CaCl_2$ resulted in the highest mechanical results due to their modulus and ultimate stress compared to other coaxial fibres.

3.3.6　Swelling Properties in SBF

The swelling properties of the fibres were determined in SBF medium over a period of 48 h. Fibre diameters were measured at different time intervals. Results are shown in Fig. 3.7 and listed in Table 3.2. Solid fibres have shown quite different degrees of swelling; while chitosan fibres showed only ~90% calcium alginate fibre, the swelling of alginate fibres occur quite fast up to high ratios (until the fibre lose its fibrillar shape completely which make it almost impossible to be measured). This phenomenon is mostly due to the ionic exchange between the divalent cations and sodium in the environment [80].

It can be seen in Fig. 3.7 that coaxial fibres containing 0.5% (w v^{-1}) calcium chloride have shown the least amount of initial swelling, while fibres with 1% (w v^{-1}) calcium chloride in the core demonstrated the highest degree of swelling. It seems that two simultaneous events are occurring by increasing the calcium chloride content in the core [from 0.5 to 2% (w v^{-1})]. Increasing the number of ionic groups (Ca^{2+}) in hydrogels is known to increase their swelling capacity [81]. This is mainly due to the simultaneous increase in the number of counterions inside the gel, which produces an additional osmotic pressure that swells the gel as described in Flory theory previously [82]. Therefore, by adding more calcium chloride to the chitosan solution the degree of swelling increases. On the other hand, increasing the amount of Ca^{2+} ions also

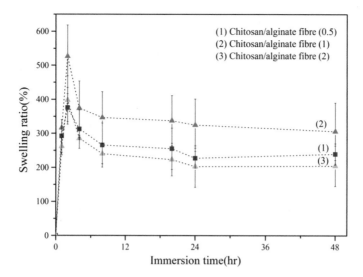

Fig. 3.7 Swelling properties of coaxial wet-spun fibres in SBF as a function of the immersion time

results in an increase in the ion exchange process within the sodium alginate. In fact, the ratio of calcium will increase in the alginate. The increase of the crosslinking agent concentration leads to the formation of a hydrogel with a greater 3D network density and so results in sheaths which show less swelling.

3.3.7 Thermogravimetric Analysis

TGA was employed as a method to determine the weight ratio of each hydrogel involved in forming the coaxial fibres based on the decomposition temperature of both materials. Thermogravimetric measurements were studied in the temperature range of 25–800 °C at a heating rate of 2 °C min^{-1} under air atmosphere. Thermal properties of as-prepared fibres are shown in Fig. 3.8.

As depicted in Fig. 3.8a, both chitosan and alginate showed almost the similar thermal behaviour below 300 °C. Alginate showed two main drops while heating at about 220 and 540 °C. The salt decomposed by dehydration followed by degradation to Na_2CO_3 at around 220 °C that decomposes in N_2 and there was also a carbonized material that decomposes slowly at about 540 °C in the air. According to the study published by Newkirk et al. [83] the Na_2CO_3 decomposition is dependent on the sample holder and atmosphere used. They reported that under N_2 the dehydration occurs in a similar way, but the decomposition of the carbonized material is slower at around 750 °C, resulting in Na_2CO_3 [82]. Two weight losses are also observed in the TGA curve of chitosan. The first stage occurs at about 270 °C with a corresponding weight loss of about 52% which is attributed to the decomposition of chitosan. The

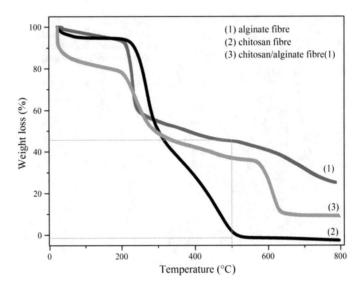

Fig. 3.8 TG curves of alginate, chitosan and coaxial Chit-Alg (1) fibres

second weight loss is around 460 °C, where a carbonized material will be decomposed in the air. The total weight loss of alginate sample at about 800 °C is 70%, while it is more than 98% for chitosan. Looking at Fig. 3.8, coaxial fibres had thermal properties and total weight loss occurring between the values observed for chitosan and alginate. At around 500 °C, only a negligible weight percentage of the chitosan fibre has been left. Consequently, the remained ingredients at this temperature are mainly from the alginate fibre which is about 28–40% wt. for the coaxial structures. These results indicate that the coaxial structures contain about 60–70% wt. of the chitosan core and 30–40% wt. of alginate sheath when dried completely. A vigorous liberation of CO_2 was observed while a dark insoluble residue remained in the test tube, in both cases.

3.3.8 Cytocompatibility Experiment

All tissue collection and handling in this chapter of my thesis was performed according to St. Vincent's Hospital approved protocols by Dr. Anita Quigley as noted previously. In order to evaluate whether the wet-spinning process induced cytotoxicity to the coaxial fibres, two types of cells were seeded on the fibres. Both human and murine myoblasts seeded onto fibre surfaces were shown to attach and align with the morphology of the fibre surface features (Fig. 3.9a–c).

Cell attachment was observed on all fibre formulations analysed in this study. Calcein staining (Fig. 3.9c) indicated that cells remained viable for at least 7 days on

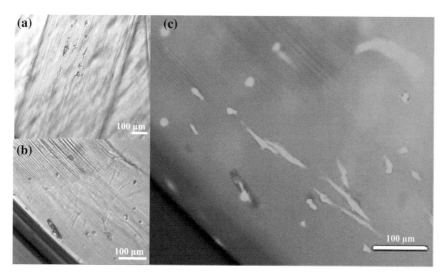

Fig. 3.9 Coaxial chitosan-alginate fibres were analysed for their ability to support primary cell adhesion and growth. Murine (**a**) and human (**b**) myoblasts were observed to adhere and spread along the alginate surface of the fibres. **c** Calcein AM staining of human myoblasts revealed that the majority of cells remained viable for at least 7 days in culture and cells appeared to show some alignment with the surface features of the fibres

fibre surfaces and attached cells generally showed normal myoblast morphology on the fibre surface, although some cells remained rounded. Cells attached to the surface of the tissue culture well showed normal morphology suggesting that the effect of the fibres on cell activity was minimal. These results suggest that alginate-chitosan fibres fabricated by the wet-spinning process showed no toxicity affecting the cell survival. Further in vivo and in vitro studies, beyond the scope of the current study, are necessary to ascertain the effects of the fibres on cell proliferation and survival and how well these fibres are tolerated in vivo.

3.3.9 In Vitro Release Measurement

The calibration curve was determined by monitoring the absorption of TB at its λ_{max} (630 nm) in SBF with various concentrations of TB using UV–vis spectroscopy. The ability of drug to release from the polymer matrix depend on a number of factors such as the solubility of the drug in the polymer matrix, the solubility of the drug in the medium, swelling and solubility of the polymer matrix in the medium and the diffusion of the drug from the polymer matrix to the medium [84]. The release profiles of TB from dye loaded coaxial fibres in SBF for up to 4 days were plotted versus time and are demonstrated in Fig. 3.10.

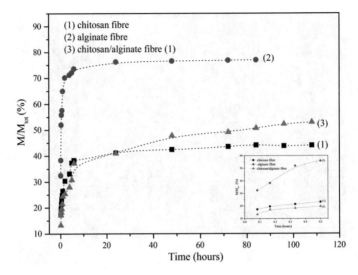

Fig. 3.10 Time dependent TB releasing behaviour of chitosan, alginate and Chit/Alg hydrogel fibres in SBF at 37 °C. Inset; burst release of coaxial fibres in the first 30 min

The whole release time period varied for different types of fibres including alginate, chitosan and the core-sheath fibre depending on the period over which they could resist the media before their structure fell apart. As noted previously, calcium alginate could be easily degraded when used for in vivo applications due to the ionic exchange between the divalent cations and sodium in the which are present in the body environment [5, 85]. Therefore, it is believed that the release of TB observed from alginate fibres was mainly due to the degradation of alginate fibres. On the other hand, wet-spinning of chitosan fibre is needed to be done in basic coagulation bath which is not an appropriate condition for most of the loaded drugs. Coaxial fibres indicated a controlled manner of release more or less like chitosan fibres. However, with the help of coaxial spinning, their fabrication process *via* wet-spinning is performed in a neutral coagulation bath. These results provide the suitable condition to load any types of drugs into the wet-spun fibres for drug delivery applications. As can be seen in Fig. 3.10, the coaxial fibres showed similar release behaviour to that of the chitosan fibres. But, they could stand the media in a shorter period of time without losing the initial structure.

In the initial period of 2 h, a fast release of TB from alginate fibres is observed at which more than 70% of TB is released. Either chitosan or Chit/Alg coaxial fibres showed approximately 30% burst release of TB followed by a sustained release within over 4 days. Whilst alginate fibre could not stand the media for more than 4 days, *ca.* 42% and 50% of the TB has released from chitosan and Chit/Alg fibres, respectively. Figure 3.10 shows a good sustained-release profile of TB from coaxial fibres. TB is a hydrophilic molecule with a greater solubility in an aqueous environment, so its drug diffusion rate through the polymeric matrix is highly dependent on the swelling of the polymeric fibre. Thus, according to the swelling ratio results, it is expected to obtain much faster release from alginate fibres than those of either chitosan or coaxial fibres.

3.4 Conclusion

The production of coaxial biofibres has been successfully developed for the first time using a wet-spinning method. The morphological, mechanical, thermal and swelling properties of these fibres are discussed. Enhanced mechanical properties of 260% in ultimate stress and more than 300% in Young's modulus were observed by incorporating 1% (w v^{-1}) CaCl$_2$ into the chitosan core. SEM micrographs of the cross-section of chitosan-alginate fibres clearly show the cylinder shaped monofilament form of the chitosan fibre covered with alginate. These biofibres as delivery platforms have demonstrated great potentials toward advancing current drug delivery systems. Hybrid Chit/Alg fibres could likely be promising as a novel kind of 3D bioscaffolds in drug release studies or tissue engineering.

References

1. Elahi F, Lu W, Guoping G, Khan F (2013) Core-shell fibers for biomedical applications—a review. J Bioeng Biomed Sci 3:1–14. https://doi.org/10.4172/2155-9538.1000121
2. Abbas AA, Lee SY, Selvaratnam L, Yusof N, Kamarul T (2008) Porous PVA-chitosan based hydrogel as an extracellular matrix scaffold for cartilage regeneration. Eur Cell Mater 16:50–51
3. Ang TH, Sultana FSA, Hutmacher DW, Wong YS, Fuh JYH, Mo XM, Loh HT, Burdet E, Teoh SH (2002) Fabrication of 3D chitosan—hydroxyapatite scaffolds using a robotic dispensing system. Mater Sci Eng, C 20:35–42
4. Breyner NM, Zonari AA, Carvalho JL, Gomide VS, Gomes D, Góes AM (2011) Cartilage tissue engineering using mesenchymal stem cells and 3D chitosan scaffolds—in vitro and in vivo assays. In: Biomaterials Science and Engineering. InTech Published, Institute of Biologic Science, Department of Biochemistry and Immunology, Brazil, pp 211–226
5. Draget KI, Smidsrùd PO, Skjåk-brñk PG (2005) Alginates from algae. In: Polysccharides and Polyamides in the Food Industry. Properties, Production and Patents. Wiley-VCH Verlag GmbH & Co. KGaA, Weinheim, pp 1–30
6. Han C, Zhang L, Sun J, Shi H, Zhou J, Gao C (2010) Application of collagen-chitosan/ fibrin glue asymmetric scaffolds in skin tissue engineering. Biomed & Biotechnol 11:524–530. https://doi.org/10.1631/jzus.B0900400
7. Iqbal M, Xiaoxue SÆ (2009) A review on biodegradable polymeric materials for bone tissue engineering applications. J Mater Sci 44:5713–5724. https://doi.org/10.1007/s10853-009-3770-7
8. Kimura Y, Hokugo A, Takamoto T, Tabata Y, Kurosawa H (2008) anterior cruciate ligament regeneration by biodegradable scaffold combined with local controlled release of basic fibroblast growth factor and collagen wrapping. Tissue Eng 14:47–57. https://doi.org/10.1089/tec.2007.0286
9. Ma L, Gao C, Mao Z, Zhou J, Shen J (2003) Collagen/chitosan porous scaffolds with improved biostability for skin tissue engineering. Biomater 24:4833–4841. https://doi.org/10.1016/S0142-9612(03)00374-0

10. Martins A, Reis RL, Neves NM (2007) Electrospun nanostructured scaffolds for tissue engineering applications. Nanomedi 2:929–942
11. Moshaverinia A, Ansari S, Chen C, Xu X, Akiyama K, Snead ML, Zadeh HH, Shi S (2013) Biomaterials co-encapsulation of anti-BMP2 monoclonal antibody and mesenchymal stem cells in alginate microspheres for bone tissue engineering. Biomater 34:6572–6579. https://doi.org/10.1016/j.biomaterials.2013.05.048
12. Seo S, Choi Y, Akaike T, Higuchi A, Cho C-S (2006) Alginate/galactosylated chitosan/heparin scaffold as a new synthetic extracellular matrix for hepatocytes. Tissue Eng 12:33–44
13. Wang T, Wang I, Lu J, Young T (2012) Novel chitosan-polycaprolactone blends as potential scaffold and carrier for corneal endothelial transplantation. Mol Vis 18:255–264
14. Wang X, Yan Y, Xiong Z, Lin F, Wu R, Zhang R, Lu Q (2005) Preparation and evaluation of ammonia-treated collagen/chitosan matrices for liver tissue engineering. J Biomed Mater Res B Appl Biomater 75B:91–98. https://doi.org/10.1002/jbm.b.30264
15. Zhang T, Wan LQ, Xiong Z, Marsano A, Maidhof R, Park M, Yan Y, Vunjak-novakovic G (2012) Channelled scaffolds for engineering myocardium with mechanical stimulation. J Tissue Eng Regen Med 6:748–756. https://doi.org/10.1002/term
16. Zhu C, Fan D, Duan Z, Xue W, Shang L, Chen F, Luo Y (2009) Initial investigation of novel human-like collagen/chitosan scaffold for vascular tissue engineering. J Biomed Mater Res A 89:829–840. https://doi.org/10.1002/jbm.a.32256
17. El-Tahlawy K, Hudson S (2006) Chitosan: aspects of fiber spinnability. J Appl Polym Sci 100:1162–1168. https://doi.org/10.1002/app.23201
18. Cascone MG, Barbani N, Cristallini C, Ciardelli G, Lazzeri L (2014) Bioartificial polymeric materials based on polysaccharides. Biomater Sci, Polym Ed 12:267–281. https://doi.org/10.1163/156856201750180807
19. Jayakumar R, Prabaharan M, Kumar P, Nair S, Tamura H (2011) Biomaterials based on chitin and chitosan in wound dressing applications. Biotechnol Adv 29:322–337. https://doi.org/10.1016/j.biotechadv.2011.01.005
20. Jayakumar R, Prabaharan M, Kumar PTS (1990) Novel chitin and chitosan materials in wound dressing. In: Biomedical engineering, trends in materials science. InTech, Amrita Centre for Nanosciences and Molecular Medicine, India, pp 3–25
21. Kucharska M, Struszczyk MH, Cichecka M, Brzoza-malczewska K (2011) Preliminary studies on the usable properties of innovative wound dressings. Prog Chem Appl Chitin Its Deriv XVI:131–138
22. Wang L, Khor E, Wee A, Lim LY (2002) Chitosan-alginate PEC membrane as a wound dressing: assessment of incisional wound healing. J Biomed Mater Res 63:610–618. https://doi.org/10.1002/jbm.10382
23. Suzuki Y, Nishimura Y, Tanihara M, Suzuki K, Kitahara AK, Yamawaki Y, Nakamura T, Kakimaru Y (1998) Development of alginate gel dressing. J Artif Organ 1:28–32
24. Jagur-grodzinski J (2006) Polymers for tissue engineering, medical devices, and regenerative medicine. Concise general review of recent studies. Polym Adv Technol 17:395–418. https://doi.org/10.1002/pat
25. Kuo CK, Ma PX (2001) Ionically crosslinked alginate hydrogels as scaffolds for tissue engineering: part 1. Structure, gelation rate and mechanical properties. Biomateri 22:511–521
26. Martel-estrada SA, Martínez-pérez CA, Chacón-nava JG, García-casillas PE, Olivas-armendariz I (2010) Synthesis and thermo-physical properties of chitosan/poly (dl-lactide-co-glycolide) composites prepared by thermally induced phase separation. Carbohydr Polym 81:775–783. https://doi.org/10.1016/j.carbpol.2010.03.032
27. Nwe N, Furuike T, Tamura H (2009) The mechanical and biological properties of chitosan scaffolds for tissue regeneration templates are significantly enhanced by chitosan from Gongronella butleri. Materials (Basel) 2:374–398. https://doi.org/10.3390/ma2020374
28. Pallela R, Venkatesan J, Janapala VR, Kim S (2011) Biophysicochemical evaluation of chitosan-hydroxyapatite-marine sponge collagen composite for bone tissue engineering. Wiley Period Inc 100:486–495. https://doi.org/10.1002/jbm.a.33292

29. Salimi A, Ghollasi M, Saki N, Rahim F, Dehghanifard A, Sh A, Hagh M, Soleimani M (2010) Application of nanoscaffolds and mesenchymal stem cells in tissue engineering. IJBC 3:11–20

30. Shao X, Hunter CJC (2007) Developing an alginate/chitosan hybrid fiber scaffold for annulus fibrosus cells. J Biomed Mater Res A 82:702–710. https://doi.org/10.1002/jbm.a

31. Steward AJ, Liu Y, Wagner DR (2011) Engineering cell attachments to scaffolds in cartilage tissue engineering. Biomater Regen Med 63:74–82

32. Vrana NE, Grady AO, Kay E, Cahill PA, Mcguinness GB (2009) Cell encapsulation within PVA-based hydrogels via freeze-thawing: a one-step scaffold formation and cell storage technique. J Tissue Eng Regen Med 3:567–572. https://doi.org/10.1002/term

33. Young C, Rekha PD, Lai W, Arun AB (2006) Encapsulation of plant growth-promoting bacteria in alginate beads enriched with humic acid. Wiley Period Inc 95:76–83. https://doi.org/10.10 02/bit

34. Belščak-cvitanović A, Stojanović R, Manojlović V, Komes D, Juranović I, Nedović V, Bugarski B (2011) Encapsulation of polyphenolic antioxidants from medicinal plant extracts in alginate—chitosan system enhanced with ascorbic acid by electrostatic extrusion. Food Res Int 44:1094–1101. https://doi.org/10.1016/j.foodres.2011.03.030

35. Bhavan M, Nagar G (2011) Chitosan—sodium alginate nanocomposites blended with cloisite 30b as a novel drug delivery system for anticancer drug curcumin Vijay Kumar Malesu, Debasish Sahoo and P. L. Nayak * ISSN 0976–4550 materials preparation of chitosan-alginate nanocomposite. Int J Appl Biol Pharm Technol 2:402–411

36. Desai KGH, Park HJ (2005) Preparation and characterization of drug-loaded chitosan-tripolyphosphate microspheres by spray drying. Drug Dev Res 128:114–128. https://doi.org/10.1002/ddr.10416

37. Rajeshkumar S, Venkatesan C, Sarathi M, Sarathbabu V, Thomas J, Basha KA, Hameed ASS (2009) Fish & shellfish immunology oral delivery of DNA construct using chitosan nanoparticles to protect the shrimp from white spot syndrome virus (WSSV). Fish Shellfish Immunol 26:429–437. https://doi.org/10.1016/j.fsi.2009.01.003

38. Rao KSVK, Kumar BV, Subha MCS, Sairam M, Aminabhavi TM (2006) Novel chitosan-based pH-sensitive interpenetrating network microgels for the controlled release of cefadroxil q. Carbohydr Polym 66:333–344. https://doi.org/10.1016/j.carbpol.2006.03.025

39. Rieux A, Duhem N, Je C (2011) Chitosan and chitosan derivatives in drug delivery and tissue engineering. Adv Polym Sci 244:19–44. https://doi.org/10.1007/12_2011_137

40. Sarma SJ, Pakshirajan K, Mahanty B (2011) Chitosan-coated alginate—polyvinyl alcohol beads for encapsulation of silicone oil containing pyrene: a novel method for biodegradation of polycyclic aromatic hydrocarbons. J Chem Technol Biotechnol 86:266–272. https://doi.org/10.1002/jctb.2513

41. Machluf M (2006) Alginate—chitosan complex coacervation for cell encapsulation: effect on mechanical properties and on long-term viability. Biopolym 82:570–579. https://doi.org/10.1002/bip

42. El-hefian EA, Nasef MM, Yahaya AH (2011) Chitosan physical forms: a short review. Aust J Basic Appl Sci 5:670–677

43. Godbey WT, Hindy BSS, Sherman ME, Atala A (2004) A novel use of centrifugal force for cell seeding into porous scaffolds. Biomaterials 25:2799–2805. https://doi.org/10.1016/j.biomaterials.2003.09.056

44. Oh SH, Lee JH (2013) Hydrophilization of synthetic biodegradable polymer scaffolds for improved cell/ tissue compatibility. Biomed Mater 8:1–16. https://doi.org/10.1088/1748-6041/8/1/014101

45. Yao J, Tao SL, Young MJ (2011) Synthetic polymer scaffolds for stem cell transplantation in retinal tissue engineering. Polym J 3:899–914. https://doi.org/10.3390/polym3020899

46. Tangsadthakun C, Kanokpanont S, Sanchavanakit N, Banaprasert T, Damrongsakkul S (2006) Properties of collagen/chitosan scaffolds for skin tissue engineering fabrication of collagen/chitosan scaffolds. J Met Mater Miner 16:37–44

47. Hussain A, Collins G, Yip D, Cho CH (2012) Functional 3-D cardiac co-culture model using bioactive chitosan nanofiber scaffolds. Biotech Bioeng 110:1–11. https://doi.org/10.1002/bit.24727

48. Koo S, Ahn SJ, Hao Z, Wang JC, Yim EK (2011) Human corneal keratocyte response to micro- and nano-gratings on chitosan and PDMS. Cell Mol Bioeng 4:399–410. https://doi.org/10.10 07/s12195-011-0186-7
49. Dutta P, Rinki K, Dutta J (2011) Chitosan: a promising biomaterial for tissue engineering scaffolds. Chit Biomater II 244:45–80. https://doi.org/10.1007/12_2011_112
50. Wang L, Li C, Chen Y, Dong S, Chen X, Zhou Y (2013) Poly (lactic-co-glycolic) acid/nanohydroxyapatite scaffold containing chitosan microspheres with adrenomedullin delivery for modulation activity of osteoblasts and vascular endothelial cells. Biomed Res Int 2013:1–13
51. Malafaya PB, Silva GA, Reis RL (2007) Natural-origin polymers as carriers and scaffolds for biomolecules and cell delivery in tissue engineering applications. Adv drug Del Rev 59:207–233. https://doi.org/10.1016/j.addr.2007.03.012
52. Quigley AF, Bulluss KJ, Kyratzis ILB, Gilmore K, Mysore T, Schirmer KSU, Kennedy EL, O'Shea M, Truong YB, Edwards SL, Peeters G, Herwig P, Razal JM, Campbell TE, Lowes KN, Higgins MJ, Moulton SE, Murphy MA, Cook MJ, Clark GM, Wallace GG, Kapsa RMI (2013) Engineering a multimodal nerve conduit for repair of injured peripheral nerve. J Neural Eng 10:1–17. https://doi.org/10.1088/1741-2560/10/1/016008
53. Costa-Pinto A, Reis R, Neves N (2011) Scaffolds based bone tissue engineering: the role of chitosan. Tissue Eng B 17:331–347. https://doi.org/10.1089/ten.teb.2010.0704
54. Nair LS, Laurencin CT (2007) Biodegradable polymers as biomaterials. Prog Polym Sci 32:762–798. https://doi.org/10.1016/j.progpolymsci.2007.05.017
55. Kumar M (1999) Chitin and chitosan fibres: a review. Bull Mater Sci 22:905–915
56. Bansal V, Sharma P, Sharma N (2011) Applications of chitosan and chitosan derivatives in drug delivery. Advan Biol Res 5:28–37
57. Qin Y (2004) Gel swelling properties of alginate fibers. J Appl Polym Sci 91:2–6
58. Qin Y (2008) The gel swelling properties of alginate fibers and their applications in wound management. Polym Adv Technol 19:6–14. https://doi.org/10.1002/pat.960
59. Dash M, Chiellini F, Ottenbrite RM, Chiellini E (2011) Chitosan—A versatile semi-synthetic polymer in biomedical applications. Prog Polym Sci 36:981–1014. https://doi.org/10.1016/j.p rogpolymsci.2011.02.001
60. Hirano S, Bash E (2001) wet-spinning and applications of functional fibers based on chitin and chitosan. Macromol Symp 168:21–30. https://doi.org/10.1017/CBO9781107415324.004
61. Khor E, Yong L (2003) Implantable applications of chitin and chitosan. Biomaterials 24:2339–2349. https://doi.org/10.1016/S0142-9612(03)00026-7
62. Lee KY, Mooney DJ (2012) Alginate: properties and biomedical applications. Prog Polym Sci 37:106–126. https://doi.org/10.1016/j.progpolymsci.2011.06.003
63. Song R, Xue R, He L, Liu Y, Xiao Q (2008) The structure and properties of chitosan/polyethylene glycol/silica ternary hybrid organic-inorganic films. Chin J Polym Sci 26:621–630
64. Xie H, Zhang S, Li S (2006) Chitin and chitosan dissolved in ionic liquids as reversible sorbents of CO_2. Green Chem 8:630–633. https://doi.org/10.1039/b517297g
65. Iwasaki N, Yamane S, Majima T, Kasahara Y (2004) Feasibility of polysaccharide hybrid materials for scaffolds in cartilage tissue engineering: evaluation of chondrocyte adhesion to polyion complex fibers prepared from alginate and chitosan. Biomacromol 5:828–833
66. Chang J, Lee Y-H, Wu M, Yang M-C, Chien C (2012) Preparation of electrospun alginate fibers with chitosan sheath. Carbohydr Polym 87:2357–2361. https://doi.org/10.1016/j.carbpol.201 1.10.054
67. Wang J, Huang X, Xiao J, Li N, Yu W, Wang W, Xie W, Ma X, Teng Y (2010) Spray-spinning: a novel method for making alginate/chitosan fibrous scaffold. J Mater Sci 21:497–506. https:// doi.org/10.1007/s10856-009-3867-1
68. Niekraszewicz A, Ciechańska D (2006) Research into the process of manufacturing alginate-chitosan fibres. Fiber Text East 14:25–31
69. Konop A, Colby R (1999) Polyelectrolyte charge effects on solution viscosity of poly(acrylic acid). Macromolecules 32:2803–2805

70. Todaro M, Quigley A, Kita M, Chin J, Lowes K, Kornberg AJ, Cook MJ, Kapsa R (2007) Effective detection of corrected dystrophin loci in mdx mouse myogenic precursors. Human 28:1–3. https://doi.org/10.1002/humu

71. Gulrez SKH, Al-Assaf S, Phillips GO (2003) Hydrogels: methods of preparation, characterisation and applications. In: Progress in molecular and environmental bioengineering—from analysis and modeling to technology applications. InTech, Glyn O Phillips Hydrocolloids Research Centre, Glyndwr University, Wrexham, United Kingdom, pp 117–149

72. Baburaj MS, Aravindakumar CT, Sreedhanya S, Thomas AP, Aravind UK (2012) Treatment of model textile effluents with PAA/CHI and PAA/PEI composite membranes. Desalination 288:72–79. https://doi.org/10.1016/j.desal.2011.12.015

73. Ding C, Xu S, Wang J, Liu Y, Chen P, Feng S (2012) Controlled loading and release of methylene blue in layer-by-layer assembled polyelectrolyte films. Mater Sci Eng, C 32:670–673. https://doi.org/10.1016/j.msec.2012.01.005

74. Paulino AT, Guilherme MR, Reis AV, Campese GM, Muniz EC, Nozaki J (2006) Removal of methylene blue dye from an aqueous media using superabsorbent hydrogel supported on modified polysaccharide. J Colloid Interface Sci 301:55–62. https://doi.org/10.1016/j.jcis.2006.04.036

75. Wang X, Zhang L, Wang L, Sun J, Shen J (2010) Layer-by-layer assembled polyampholyte microgel films for simultaneous release of anionic and cationic molecules. Langmuir 26:8187–8194. https://doi.org/10.1021/la904558h

76. Dragan ES, Felicia D, Loghin A (2013) Enhanced sorption of methylene blue from aqueous solutions by semi-IPN composite cryogels with anionically modified potato starch entrapped in PAAm matrix. Chem Eng J 234:211–222

77. Barroso T, Soares T, Beatriz C, De Paula C, Bernadete M, Pierre R, Aparecida S, Pereira DL, Massao M, Cristina R, Teresa M, Garcia J (2015) Using chitosan gels as a toluidine blue O delivery system for photodynamic therapy of buccal cancer: in vitro and in vivo studies. Photodiagnosis Photodyn Ther 12:98–107

78. Tronci G, Sri R, Arafat MT, Yin J, Wood DJ, Russell SJ (2015) Wet-spinnability and crosslinked fibre properties of two collagen polypeptides with varied molecular weight. Int J Biol Macromol 81:112–120

79. Cong H-P, Ren X-C, Wang P, Yu S-H (2012) Wet-spinning assembly of continuous, neat, and macroscopic graphene fibers. Sci Rep 2:613. https://doi.org/10.1038/srep00613

80. Qin Y (2005) Ion-exchange properties of alginate fibers. Text Res J 75:165–168. https://doi.org/10.1177/0040517505075002 14

81. Okay O, Durmaz S (2002) Charge density dependence of elastic modulus of strong polyelectrolyte hydrogels. Polym J 43:1215–1221

82. Pillai CKS, Paul W, Sharma CP (2009) Chitin and chitosan polymers: Chemistry, solubility and fiber formation. Prog Polym Sci 34:641–678. https://doi.org/10.1016/j.progpolymsci.2009.04.001

83. Newkirk AE, Aliferis I (1958) Drying and decomposition of sodium carbonate. Anal Chem 30(5): 982–984

84. Phu G (2007) Factors influencing drug dissolution characteristic from hydrophilic polymer matrix tablet. Sci Pharm 75:147–163

85. Davidovich-pinhas M, Bianco-peled H (2010) A quantitative analysis of alginate swelling. Carbohydr Polym 79:1020–1027. https://doi.org/10.1016/j.carbpol.2009.10.036

Chapter 4
Fabrication of Coaxial Wet-Spun Biofibres Containing Graphene Core

4.1 Introduction

Electrical stimulation has been able to provide beneficial effects for regeneration of tissue: muscle [1–4], nerve [5, 6] and bone [7]. A range of metallic electrodes, most commonly Pt or Ti based, have conventionally been used [4, 8]. More recently the use of organic conductors has attracted attention since they can be loaded with bioactivity enhancing the electro-cellular communication process [9, 10]. Both organic conducting polymers [11–13] and graphene [14–16] have proven to be useful in this regard. However, in both cases, the mechanical properties necessary to engender adequate conductivity are not an ideal match with those of the surrounding tissue. Formation of coaxial structures, wherein the conductor is surrounded with a more cyto-friendly material, should provide alternate fabrication options when considered as the building blocks of 3D structures. Use of a coaxial structure is an excellent alternative which allows improved characteristics when compared to those of the single component fibres and has the added advantage of providing a flexible system which may be optimised for a variety of purposes. Coaxial spinning is an advanced processing technique that has just been used recently to produce hybrid fibres from natural and synthetic polymers [17] for bioapplications. Using this technique, it would become possible to take the advantages of an electroactive core for electrical stimulation purposes inside the body as well as providing good biocompatibility from the hydrogel-based sheath for cell adhesion and differentiation [18].

To date, several previous studies have used coaxial electrospinning which is the simplest approach in the field of coaxial spinning [19]. Wet-spinning has been used to produce fibres with similar structures to that of coaxial fibres (detailed in

Parts of this chapter have been reproduced with permission from https://doi.org/10.1002/adem.20 1500201 (Mirabedini, A., Foroughi, J., Thompson, B. and Wallace, G. G. (2016), Fabrication of Coaxial Wet-Spun Graphene–Chitosan Biofibres. Advanced Engineering Materials, Volume 18: pp. 284–293).

© Springer Nature Switzerland AG 2018
A. Mirabedini, *Developing Novel Spinning Methods to Fabricate Continuous Multifunctional Fibres for Bioapplications*, Springer Theses,
https://doi.org/10.1007/978-3-319-95378-6_4

Sect. 1.3.3) [20–23]. There are only a few reports in the literature of fabricating coaxial fibres using coaxial spinnerets for wet-spinning. Recently, Liang Kou et al. have also reported on the production of CMC/wrapped graphene/CNT coaxial fibres for supercapacitor applications [24]. The key principle in wet-spinning is to make polymer fibres by a transition from a soluble to a non-soluble phase [25]. The major difference of conventional wet-spinning with the coaxial method is that in the coaxial process, two different polymer solutions are injected into a coaxial spinneret together and are coagulated in a bath to form a fibre retaining a coaxial structure.

Chitosan is a naturally occurring aminopolysaccharide, which has recently generated great interest for its potential in clinical and biological applications [26–29]. It can be regarded as a derivative of cellulose in which the C-2 hydroxyl group in anhydroglucose units of cellulose is replaced by a free amine group [30]. Alginate is a linear, binary copolymer composed of 1, 4-linked β-D-mannuronic acid (M) and α-L-guluronic acid (G) monomers. The composition of alginate (the ratio of the two uronic acids and their sequential arrangements) varies with the source. Salts of alginic acid with monovalent cations such as sodium alginate are all soluble in water.

Graphene was selected as the conductive core constituent. Graphene is a two-dimensional atomically thin mesh of carbon atoms which has recently received widespread attention. Numerous outstanding properties are exhibited by this material such as its excellent conductance of both heat and electricity, and being the strongest material ever measured [31, 32]. Despite the properties of aqueous reduced graphene oxide (rGO) suspensions, they are not directly spinnable by themselves. This could be due to their incoherent particles, low viscosity and the high tendency of graphene sheets, which have a high specific surface area, to form irreversible aggregation through van der Waals interactions [33]. However, spinning of liquid crystalline (LC) suspensions of large sheet graphene oxide (GO) in water has been recently reported by several research groups [34, 35]. Use of large GO sheets has enabled the use of a wet-spinning route to produce strong fibres which can be easily converted to electrically conducting graphene fibres by using an appropriate chemical reducing agent. L-Ascorbic acid (L-AA) is a non-toxic naturally occurring compound known to act as a reducing agent in physiological processes and has also been used as a primary reducing agent in the laboratory [36]. In addition, It has been reported earlier that the epoxy groups in GO can be crosslinked easily to the primary amine groups in chitosan, and so the polysaccharide has been widely used to modify GO [37]. Using this interaction as well as the extensive H-bonding between the polymer and GO, it is possible to form a homogenous robust connection between chitosan and GO with favourable properties for coaxial wet-spinning.

The aim of current study is to establish a wet-spinning process to produce coaxial conductive biofibres to take the benefit from the electroactivity of the conductive core along with improving the mechanical properties of the fibres and keeping good biocompatibility and cell adhesion of the scaffold by using a biomaterial for the sheath. Both GO and chitosan have been widely applied in the studies interacting with cells [16, 38, 39]. Specifically, we have developed a one-step fabrication method to produce coaxial fibres of chitosan and GO *via* a facile continuous technique. Spinnability of coaxial fibres considering spinning conditions such as appropriate choice of the hydrogel, appropriate GO concentration and chitosan viscosity, injection flow rate of each spinning solution, identification of suitable reducing agent selection and reduction method have been investigated in this study. Herein, we also present a simple approach for reducing GO fibres using L-AA as a reducing agent in an aqueous solution at 80 °C as also suggested elsewhere [40]. We discuss the maximum conductivity, an increment in mechanical properties and biocompatibility sufficient to support cultured cell attachment and growth. The preparation of for three fibre types (Alg/GO, Chit/GO and rGO) have been reported and compared.

4.2 Experimental

4.2.1 Materials

High molecular weight chitosan was used in this work to provide better mechanical properties. L-Ascorbic acid (L-AA) were purchased from Sigma-Aldrich Co. used as the reducing agents.

A simulated body fluid (SBF) solution, with ion concentrations approximately equal to those of human blood plasma, has been prepared with the following ion concentrations of 142 mM Na^+, 5 mM K^+, 1.5 mM Mg^{2+}, 2.5 mM Ca^{2+}, 103 mM Cl^-, 27 mM HCO_3^-, 1.0 mM HPO_4^{2-} and 0.5 mM SO_4^{2-} with the final adjusted pH of 7.4 ± 0.05 [41]. This solution was used as an aqueous medium to reswell the dried fibres for imaging them in wet-state.

4.2.2 Fibre Spinning

4.2.2.1 Wet-Spinning of GO

Large GO sheets were obtained using the modified Hummers method [35]. These were dispersed in deionized water only by gentle shaking, without sonication. Two types of fibre wet-spinning methods were utilized, rotary solid wet-spinning and long-bath coaxial wet-spinning. Wet-spinning of GO fibres was performed using a coagulation bath containing aqueous 2% (w v^{-1}) calcium chloride (Ethanol/H$_2$O: 1:5) using a rotary wet-spinning apparatus. The GO suspension (0.63 w v^{-1}) was injected into the coagulation bath sitting on a rotating stage (~60 rpm) at a flow rate of 15 mL h^{-1}. GO fibres were then placed into a bath of MilliQ water containing reducing agent (L-AA) with the weight ratio of 3 relative to the concentration of GO and heated in an oven to 80 °C to reduce the GO fibres to rGO. Dried fibres were obtained by washing the gel-state fibres in MilliQ water and then air-dried under a slight tension to avoid being tilted when dried at room temperature. This allows us to obtain conductive fibres with the addition of a post treatment step.

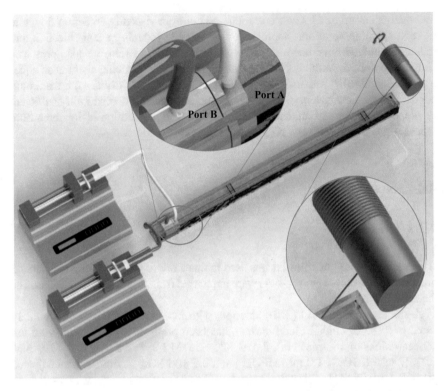

Fig. 4.1 A schematic of coaxial wet-spinning set up for producing Chit/GO coaxial fibres

4.2.2.2 Coaxial Wet-Spinning of Chitosan/GO and Alginate/GO

The gel spinning precursors were prepared as described in Sect. 2.1.2. Using the same concentration of GO suspension we used earlier to make GO fibres (0.63 w v^{-1}), coaxial fibres of Alginate/GO (Alg/GO) and chitosan/GO (Chit/GO) was then spun into a bath containing 2% w v^{-1} $CaCl_2$ and 1 M sodium hydroxide (Ethanol/H_2O: 1/5), respectively. The feeding rates of 10 and 18 mL h^{-1} were employed for GO and both gel precursors, respectively in order to provide sufficient time for the sheath component to cover the core. The coaxial wet-spinning process was carried out as described before in Sect. 2.2.1.2. The setup consists of two injection syringes and pumps (KDS100, KD Scientific Inc.), connected to the ports of the predesigned coaxial spinneret that allows for injection of two fluids, a coagulation bath and a stretching collector as illustrated schematically in Fig. 4.1. The GO spinning solution was injected as the core component and drawn through the centre outlet nozzle into the coagulation bath. At once, alginate or chitosan spinning solution was injected as the sheath of the fibre, providing an outer casing for the GO core, by feeding through port A.

Hydroxyl and amino groups from chitosan can interact with oxygen-containing groups including hydroxyl, carbonyl and carboxyl moieties in the graphene oxide sheets forming hydrogen bonding as previously studied. In addition, it is suggested that not only amine groups can take part in the reaction, also the OH, of the chitosan structure can play an important role [42].

4.2.3 Characterisations of rGO and Coaxial Fibres

4.2.3.1 Material Parameters Characterisation

Changes in viscosity have been recorded versus shear rate. The rheological properties of alginate [3% (w v^{-1})], chitosan [3% (w v^{-1})] and GO (0.63 w v^{-1}) solutions were examined in flow mode (cone and plate method) for each sample at room temperature (~25 °C) with the shear rates between 0.1 and 300 s^{-1}.

4.2.3.2 Fourier Transform Infrared

FTIR spectra of the chitosan and Chit/GO fibres were recorded to evaluate the possible reactions between core and sheath components. FTIR spectroscopy was performed in KBr pellet on a Shimadzu FTIR Prestige-21 spectrometer, in the 700–4000 cm^{-1} range with 4 cm^{-1} resolution.

4.2.3.3 Contact Angle Measurement

Contact angle measurement of distilled water on single fibres was conducted with a contact angle system OCA (dataphysics) equipped with SCA20 software. The sessile drop method, which there a sitting drop of water is resting on a fibre surface, was used at 5µL s^{-1} rate to measure advancing to the receding contact angles on already suspended fibres. Then, the optical tensiometer combined with dispenser is used to measure advancing and receding contact angles between water and fibres surface.

4.2.3.4 Raman Spectroscopy

Raman spectra were recorded on a Jobin Yvon Horiba HR800 Raman microscope using a 632 nm laser line. Individual polymer fibres were embedded in Epoxy resin (Epofix) at room temperature and polymerised overnight at 80 °C. The polymer blocks were then faced off on a Leica UC7 microtome at right angles to the fibre orientation using freshly prepared glass knives. This method resulted in a smooth block surface and relatively flat cross-sections of the individual fibres. Raman spec-

(a)   **(b)**

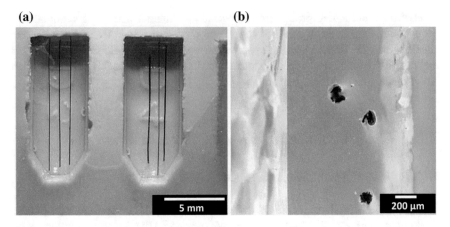

Fig. 4.2 **a** Embedded fibres in Epoxy resin at room temperature, **b** fibre cross-sections through cut surface block

troscopy was then carried out on the fibre cross-sections. The sample preparation method is demonstrated in Fig. 4.2.

Raman spectra were recorded on a Jobin Yvon Horiba HR800 Raman microscope using a 632 nm laser line and a 300-line grating to achieve a resolution of ± 1 mm^{-1}.

4.2.3.5   In Vitro Cytocompatibility Experiments

All cell culture studies in this chapter of my thesis were performed at the University of Wollongong, Australian Research Council Centre of Excellence by Dr. Brianna Thompson. After the sample preparations previously described in Sect. 2.2.1.13, the adhesion and proliferation on coaxial and rGO fibres were determined for two cell lines, a mouse fibroblast line (L-929) and a rat pheochromocytoma line commonly used as a model of neural differentiation (PC-12). L-929 cells were seeded at 5×105 cells cm-2 of culture area in DMEM supplemented with 10% fetal bovine serum and 1% penicillin/streptomycin and incubated at 37 °C and 5% CO_2 for 3 days. After the culture period, L-929 cells were live/dead stained by addition of calcein/propidium iodide (at 1 μM calcein and 1 μg mL^{-1} propidium iodide) into media, incubating for 15 min, then imaging with Leica DM IL LED microscope. Cell survival was assessed for fibroblast cells both on the fibres as well as cells growing on the tissue culture plastic underneath by analysis of confocal images. Live cells were calcein-stained (bright green) and dead cells were counted as any cell which had a brightly propidium iodide stained nucleus. PC-12 cells were seeded at 2×103 cells cm^{-2} in proliferation media (DMEM supplemented with 10% horse serum, 5% fetal bovine serum and 1% penicillin/streptomycin) and changed to differentiation media (DMEM supplemented with 2% hose serum and 1% penicillin/streptomycin) after 24 h in culture. The cells were allowed to differentiate for 10 days in culture (with a change to fresh differentiation media every 2–3 days). PC-12 cells were fixed with 3.7% paraformaldehyde and stained with Alexa-488 phalloidin to visualise the cytoskeleton. The cells were imaged using a Leica SP5 confocal microscope, with a Z-stack thickness of 1.8 µm and tiled images were combined to image along the fibre.

4.3 Results and Discussion

4.3.1 Optimization of Spinning Solutions

Initial experiments involved characterisation of alginate and chitosan solutions as well as the GO suspension. The fibre formation is dependent on the rheological properties of the spinning solution as well as appropriate coagulation chemistry [43].

4.3.1.1 Solution Concentration

Often there is an upper and lower limit for each material concentration during wet-spinning facilitates spinning continuous fabrication of fibres as discussed. At low concentrations, there is insufficient chain entanglement for forming a fibrillar structure. Although, using very high concentrations of spinning solutions can lead to a formation of non-uniform fibres. The range of reported spinnable concentrations for chitosan reported so far significantly varies from 2 to 15% (w v^{-1}) [44–47]. However, water-soluble chitosan which possesses a higher solubility threshold (up to 15% (w v^{-1}) for low molecular weight) cannot form fibres inside aqueous alkaline bath as observed. Medium molecular weight chitosan in a powder form was found to be dissolved into an aqueous acetic acid solution, not more than 5% (w v^{-1}). Still, the upper limit cannot be wet-spun in a continuous and uniform manner. The optimum concentration range for wet-spinning into a coagulation bath of 1 M sodium hydroxide was found here to be 2–5% (w v^{-1}). The mechanical properties enhance as a result of increasing the gel concentration. However, high solution viscosities due to the polyelectrolyte effect [27] hinder continuous flow of the solution into the nozzle for wet-spinning purposes as mentioned. Observations also indicated that aqueous alginate solutions at a concentration of below 2% (w v^{-1}) would not generate a continuous fibrous structure via wet-spinning; increasing the alginate concentration from 2 to 4% (w v^{-1}), the solution became highly spinnable. Then again, at concentrations above 4% (w v^{-1}), the solution became highly viscous and this impeded continuous flow through the needle, rendering the solution unspinnable. A concentration of 3% (w v^{-1}) has been selected for both gels due to the ease of spinnability, together with maintaining the suitable mechanical properties for coaxial wet-spinning as described before in our previous study [48].

In another study, researchers have investigated the rheological behaviour and spinnability of GO suspensions at concentrations ranging from 0.01 to 0.5 w v^{-1}. This study found that two different phases exist in GO suspensions at dissimilar concentrations- biphasic and nematic. These phases determined the lower and upper thresholds which GO suspensions become spinnable [35]. Below 0.025 w v^{-1}, GO suspensions were in the biphasic phase, and were completely isotropic and unspinnable. At concentrations between 0.025 and 0.075 w v^{-1}, a transition of biphasic to fully nematic phase is observed. As a consequence, suspensions were found to have partial spinnability. This partial spinnability was characterised by the weak cohesion of the GO suspension upon injecting in the coagulation bath, which resulted in short lengths of gel-state GO fibres. Using GO concentration of ≥0.075 w v^{-1}, when the fully nematic phase forms, long lengths of robust gel-state fibres could be wet-spun. GO concentrations from 0.075 to 0.5 w v^{-1} showed similar ease of spinnability [35]. Here, to improve the mechanical properties of the fibres rather than leaving an unbroken core after the fibre formation, the concentration of 0.63 w v^{-1} has been chosen.

4.3.1.2 Solution Viscosity

The rheological properties of the sheath spinning solution are critical to support the formation of a core-sheath structure formation [49]. It has previously been suggested that the requirements for the spinnability of the core solution are not as critical [50]. However, here we found that the break-up of the core was observed when the viscosity of the core dope was too low (≤0.3 Pa*s). Accordingly, the core fluid must also possess a certain minimum viscosity if it is to be entrained continuously without break-up. Figure 4.3a–c relates concentration with viscosity obtained from GO at 0.63 w v^{-1}, alginate and chitosan at 3% (w v^{-1}) at different shear rates.

It can be seen that by increasing the shear rate for all solutions, the viscosity decreased. This means that all spinning solutions exhibited shear thinning behaviour. However, the viscosity of both gel precursors was significantly higher than that of GO suspension at all shear rates. This difference reaches ~7 and ~5.5 times in the low shear rate regime for chitosan and alginate, respectively. The viscosity of the 3% (w v^{-1}) chitosan spinning solution was 10.8 Pa*s, while the viscosity of 0.63 w v^{-1} GO solution was approximately 1.5 Pa*s in low shear rates close to zero. Alginate solution of 3% (w v^{-1}) resulted in the viscosities of approximately 8.5 Pa*s.

Nevertheless, their viscosity of the core and sheath solutions was closer at higher shear rates. Considering V$_i$ for each component (10 mL h^{-1} for GO suspension and 18 mL h^{-1} for alginate and chitosan solution) and outlet sectional area, the flow rate can be calculated for both components. While spinning, the output shear rate of alginate or chitosan is ~48 and 6.81 s^{-1} for GO which is seven times more than GO (detailed in Sect. 2.2.1.2.1). Therefore, alginate and chitosan solution would have the viscosity of ~4 Pa*s during the spinning process while GO solution holds the viscosity of 1.35 Pa*s which was also shown in Fig. 4.3.

Fig. 4.3 The viscosity of spinning solution **a** chitosan 3% (w v^{-1}), **b** alginate 3% (w v^{-1}) and GO suspension (0.63 w v^{-1}) as a function of shear rate

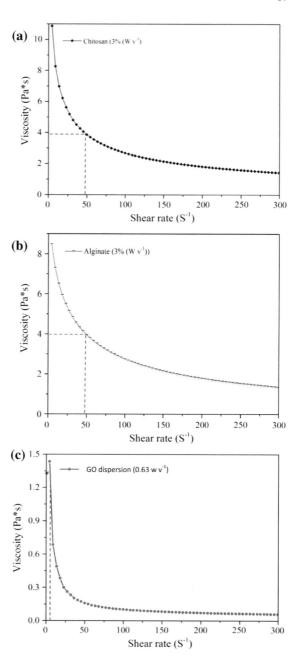

4.3.2 Morphology of As-Prepared Fibres

The internal microstructures of the coaxial fibres in both wet and dry states have been observed using LV-SEM and SEM, respectively.

4.3.2.1 Morphological Observation in Wet-State

Stereomicroscope and LV-SEM images of as-spun hydrated wet-spun rGO and coaxial Chit/GO and Alg/GO fibres are shown in Figs. 4.4 and 4.5, respectively. The coaxial fibres showed uniform and straight structures with the graphene oxide core loaded in the centre of fibres. Moreover, the core component in Chit/GO fibres showed fast, unexpected colour change (from yellow-brown to black) when exposed to alkali solutions suggesting the quick deoxygenation of graphene oxide as reported previously [51, 52]. In addition, exposing of GO to the acidic solution of the sheath component (chitosan + acetic acid) could lead to the chemical reduction of GO resulting in higher degree of hydrogen bonding within the structure as discussed elsewhere [40]. The diameters of GO, Chit/GO and Alg/GO fibres were ~100 ± 10, ~260 ± 30 and ~290 ± 30 μm, respectively.

Fibre internal structures were also observed by LV-SEM in their wet-state. The micrographs are shown in Fig. 4.5. Compositional analysis of the cross-sections of Chit/GO fibre was also carried out to detect a higher contrast between areas with different chemical compositions. In other words, composition determines the image contrast in a compositional image. Heavy elements (high atomic number) backscatter electrons more strongly than light elements (low atomic number), and thus appear brighter in the image. Therefore, backscattered electrons emitted from a sample could be captured which accounts for different types of information regarding the compositional structure of the sample [53].

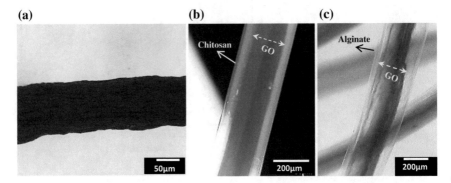

Fig. 4.4 Illustrative optical microscopic images of **a** rGO fibre, **b** coaxial Chit/GO and **c** Alg/GO coaxial fibre in wet-state

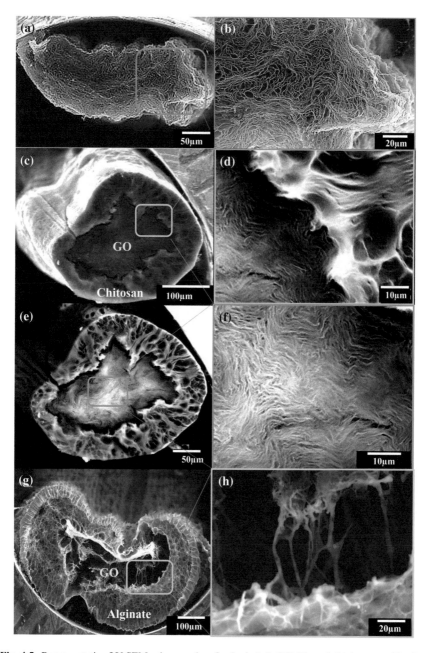

Fig. 4.5 Representative LV-SEM micrographs of **a** hydrated rGO fibres, **b** higher magnification of rGO, **c** cross-section of coaxial Chit/GO fibre, **d** higher magnification of Chit/GO interface in coaxial fibres, **e** cross-section of coaxial Chit/GO fibre in compositional mode, **f** layered morphology of graphene core with dentate bends in wet-state, **g** cross-section of coaxial Alg/GO fibre and **h** higher magnification of Alg/GO interface in coaxial fibres

In the rGO fibres, the cross-sectional areas are not perfectly circular. This phenomenon has also been reported for GO fibres elsewhere [54, 55]. In contrast, the cross-section of coaxial fibres (Fig. 4.5c) hold the round shape of the extrusion tip much more closely. It can be seen that two materials (chitosan and GO) attached entirely together. The chitosan sheath looks porous and spongy under LV-SEM, while the surface of graphene appears to shows quite dense structure with "dentate bends", which is similar to the sheet-like section structures of the pitch-based carbon fibres [55]. These dentate bends could originate from the orientational declinations (points or lines) in the spinning dopes [55] which usually formed during the dehydration of the gel fibres [56].

In contrast, GO appeared to blend with the alginate fibre component while spinning. This is why we ended up creating a nearly hollow fibre of Alg/GO. GO contains negative charge carriers on the edge of its chemical structure mainly in the form of hydroxyl, carbonyl and carboxyl functional groups as depicted schematically in Fig. 4.6 [57]. Sodium alginate [being R–COO-Na$^+$, the salt of a carboxylic acid, CH_3COOH] is also negatively charged. Lack of any chemical reactions between GO and alginate along with low viscosity of GO (~1.5 Pa*s) due to its non-polymeric nature compared to alginate (~8.5 Pa*s) can lead to a blending of these components.

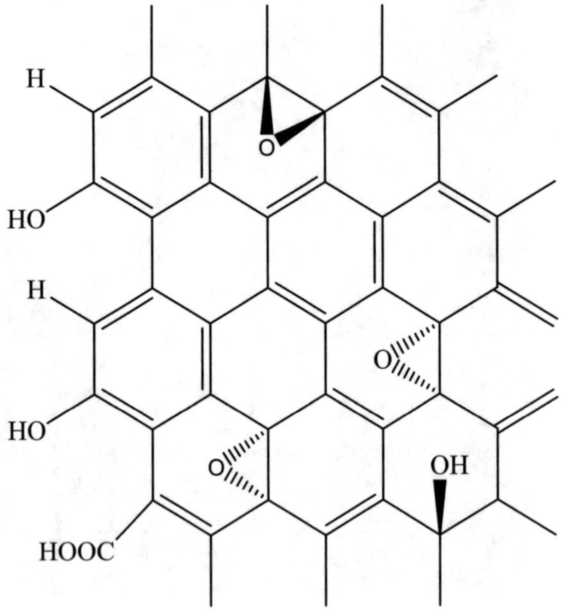

Fig. 4.6 Schematic representation of the functional groups on the edge of GO structure

4.3.2.2 Morphology of As-Prepared Fibres in Dry-State

Cross-sections of dehydrated rGO and reduced Chit/GO fibres were imaged with SEM as indicated in Fig. 4.7. It can be seen from SEM images that rGO fibres appear much more uniform in shape than in the hydrated state (Fig. 4.7a). As observed in LV-SEM micrographs previously, cross-sections of Chit/GO fibres in SEM micrographs clearly show the layered dense structure of graphene sheets which is surrounded by a thick porous chitosan sheath (~25 μm) in coaxial fibres (Fig. 4.3c). The coaxial fibres were also sufficiently flexible and tough to tie into knots (Fig. 4.3e). GO and rGO fibres could not be tied into tight knots without breaking, although the morphology of them have previously been reported to be dense and pore-free.

4.3.3 FTIR Spectroscopy Results

FTIR spectroscopy was used to evaluate the possible chemical reactions between chitosan and GO as they exposed to each other. Consequently, Fig. 4.8 shows the resulting FTIR spectra of chitosan and Chit/GO fibre.

Pure chitosan displayed characteristic absorption bands at 3352, 2878 cm^{-1}, attributed to the –OH and –CH3 groups as expected [58]. Furthermore, bands were identified at 1590 and 1404 cm^{-1} typical of the N–H group bending vibration and vibrations of –OH group of the primary alcoholic group, respectively. The bands at 1320 and 1077 cm^{-1} correspond to the stretching of C–O–N and C–O groups [59]. The band corresponding to free acetic acid (1706 cm^{-1}) was not identified. Interestingly, Chit/GO showed different peak intensities from pure chitosan, supporting the theory of the creation of a hydrogen bonding between hydroxyl and amino groups from chitosan and oxygen-containing groups including hydroxyl, carbonyl and carboxyl moieties in the graphene oxide. Subsequently, a dramatic decrease in intensity of the –OH stretching vibration (at 3221 cm^{-1}) of chitosan might be due to the reaction of some of the hydroxyl groups. In addition, both peaks at 1590 and 1404 cm^{-1} which assigned to the N–H bending of NH$_2$ groups were disappeared in Chit/GO sample followed by the emergence of peaks at 1038 cm^{-1} which corresponds to the absorbance of glucosidic bond, stretching vibration from C=O of –NHCO [60]. Significant decreasing in the intensities of O–H and N–H groups followed by formation of the new peak of C=O together with the Raman spectra results could prove the possible formation of a hydrogen bonding between chitosan and GO. Thus, not only amine groups can take part in the reaction, but also the OH, of the chitosan can play an important role [42]. It is also worth noting that the noticeable peak at 2380 cm^{-1} was due to detection of CO$_2$ gas known as the error of the device which could not be even removed by purging the interferometer with N$_2$ gas at a low flow rate for 2–3 min and performing a background before the test [61].

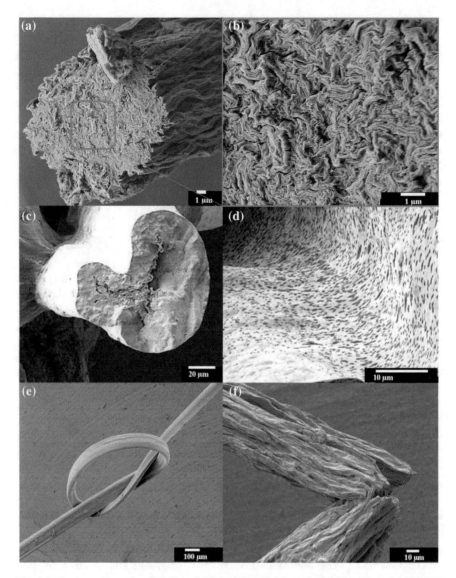

Fig. 4.7 Cross sections of **a** rGO fibre, **b** higher magnification of rGO fibre, **c** coaxial Chit/GO fibre, **d** surface of Chit/GO oxide fibre and **e** tied coaxial fibre and **f** brittle rGO fibre

4.3.4 Mechanical and Electrical Properties

Stress-strain curves were obtained for rGO, GO, chitosan and Chit/GO fibres to examine their mechanical properties. These properties tested at room temperature and the average of the four specimens tested is given in Table 4.1. A comparison

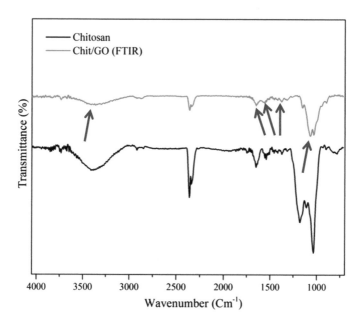

Fig. 4.8 FTIR spectra of chitosan and Chit/GO showing the possible chemical reactions between those

Table 4.1 Mechanical properties and conductivity results of GO, rGO, chitosan and Chit/GO fibres

Fibre Name	Young's Modulus (GPa)	Stress at Break (MPa)	Strain at Break (%)	Conductivity (S m^{-1})
GO	0.29±0.1	~3.2±2	1.5±0.1	19±0.1
rGO	0.34±0.1	~3.5±3	1.3±0.05	48±0.48
Chitosan	10.6±0.2	~272.3±1	8.05±0.21	–
Reduced Chit/GO	9.7±0.5	~257±5.2	3.1±0.08	–

of the mechanical properties of coaxial Chit/GO, chitosan and rGO fibres are also shown in Fig. 4.9.

Stress–strain curves obtained from as-spun rGO and coaxial Chit/GO fibres show a significant increase in robustness of the latter. Young's moduli of these rGO fibres were shown to be ∼0.29 and ∼0.34 GPa for GO and rGO fibres, respectively. These data values were enhanced about 33 times up to ~10.6 GPa for the coaxial fibres which showed similar ultimate stress values to chitosan fibres but less elongation at break. Analysis of these curves also indicated a stress at break of ∼157 MPa with ∼3.1% strain for coaxial fibres compared with only ∼3.2 and 3.5 MPa stress with ∼1.5 and ∼1.3% strain for GO and rGO fibres, respectively. In other words, there is a loss of ∼5% of tensile strain in hybrid fibres which makes the value close to that of GO fibre and a gain of ∼86% of ultimate tensile stress in the coaxial fibre. The high ultimate tensile stress achieved by coaxial fibres could be explained

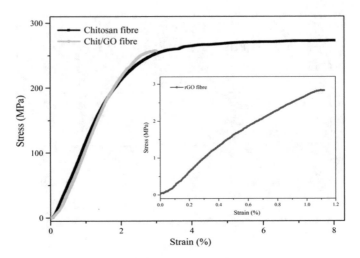

Fig. 4.9 Stress-strain curves obtained from the uniaxial tensile test on chitosan and coaxial Chit/GO fibres. Inset; Stress-strain curves obtained from the uniaxial tensile test on rGO fibres

by the creation of a strong homogenous interfacial connection between chitosan and graphene oxide which causes efficient stress transfer when strain is applied on a tensile specimen. Furthermore, it is observed from the curve that the facture of both components happened at one step in coaxial fibres. The coaxial fibre showed a fracture in the elastic regime, while chitosan fibres revealed plastic deformation prior to fracture. This elastic deformation known as a "brittle fracture" has been also observed for GO and rGO fibres. Thus, lowering the elongation at break in coaxial fibres compared to that of chitosan fibres might be the result of strong interfacial adhesion between chitosan and GO due to the hydrogen bonding. Consequently, both chitosan and Chit/GO fibres were also able to withstand higher stresses before reaching fracture. These brittleness performances of rGO fibres compared to the ductile nature of Chit/GO fibres was also observed later on in biological experiments. These data were calculated assuming that the cross-sectional area was circle with a diameter equal to the longest width (widest diameter) of the irregular fibre.

Electrical conductivity of graphene fibres (GO and rGO) was also determined. As mentioned previously, because of the presence of an insulating chitosan layer as the sheath in coaxial fibres it was not possible to measure the conductivity through 4-point probe method. The average electrical conductivity of undrawn GO fibres was measured to be $\sim$19 S m^{-1} which was improved more than 2.5 times to $\sim$48 S m^{-1} for rGO fibres.

4.3.5 Surface Wettability of As-Spun Fibres

Drop shape analysis is a convenient way to measure contact angles and thereby determine surface wettability. The quantitative measurement of liquid-solid interaction is the contact angle (θ), made by a water droplet placed against a solid. The contact angle may be related to the surface energies (γ)s of the three interface by Young's equation ($\gamma_{sv} - \gamma_{sl} = \gamma_{lv} \cos \theta$), where γ_{sv} is the surface free energy of the solid fibre in contact with air, γ_{sl} is the surface free energy of the solid fibre covered with water, γ_{lv} is the surface free energy of the water-air interface and θ is the contact angle. As it appears in Fig. 4.10, the coaxial fibres can be defined as hydrophilic, as the contact angle are ~45° (<90°). However, the contact angles indicated to be increased to 72° and 112° in chitosan and rGO fibres, respectively which are described as a more hydrophobic behaviour. It is worth mentioning that GO fibres showed almost similar θ values as that of GO fibres (~110°). Surface free energy is one of the key parameters that guides the first events occurring at the biomaterial/biological interface [62]. Therefore, the hydrophilic nature of coaxial Chit/GO fibre surface suggests that they may have potentially better interactions with biological cells.

4.3.6 Cyclic Voltammetry

CV curves of as-prepared coaxial fibres have been achieved to evaluate the EC properties of rGO and reduced Chit/GO fibres. The current measured is proportional to the number of sites that can be reduced or oxidized at a given potential, and this quantity depends not only on the chemical composition of the polymer but also on structural factors such as conjugation length and local order. Thus, they could be potentially used as an accurate tool to identify a given polymer type. CV for rGO and reduced Chit/GO fibres is shown in Fig. 4.11a, b in two aqueous electrolytes, PBS and aq. 1 M NaCl. To make a connection to the reduced graphene core of the fibres, a cotton-steel wire with the average diameter of 20 μm was inserted into the

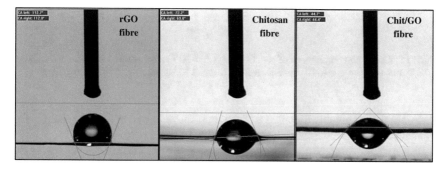

Fig. 4.10 Water contact angle measurement on suspended fibres

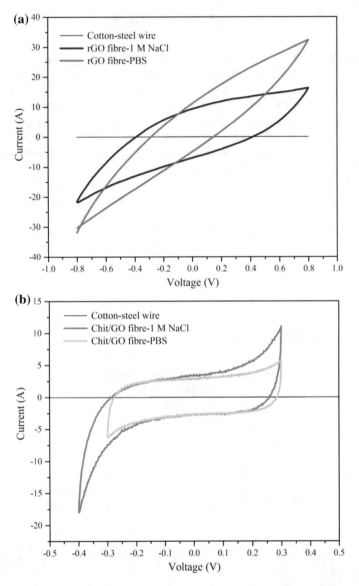

Fig. 4.11 Cyclic voltammograms of **a** rGO and **b** Chit/GO fibres; potential was scanned between −0.8 V and +0.8 (*vs.* Ag/AgCl) in PBS electrolyte and 1 M aqueous NaCl solution at 50 mVs^{-1}

fibre core while spinning (detailed and schematically showed in Sect. 2.2.1.2.1). The applied scan rate (υ) was 50 mV s^{-1} and 40 cycles were performed.

The transfer of electrons from/to graphene to/from molecules is related to the amount of defects, functional groups, and impurities present on graphene. In addition, GO prepared by different preparation methods also exhibits different reduction potentials, which might be related to the oxidizing agent used in the preparation. The EC performance of individual rGO fibres have been previously studied in a three-electrode cell using different electrolytes. Zhou et al. reported of wide EC potential window of graphene electrodes in 0.1 M PBS [63]. Graphene has been also reported as an active material for electrodes in supercapacitors exhibiting superior performance in regards to specific capacitance; for example, bioinspired solvated graphene [64] based supercapacitors gave about 215 F g^{-1}, while chemically modified graphene gave about 135 F g^{-1} [65]. Looking at the Fig. 4.9a, compared with the quite distorted shape of the CV curves of rGO fibres in PBS solution, those of the responses in 1 M NaCl were much closer to the quasi-rectangular shape, indicating the faster representative of good EC performance. The fibres showed the specific capacitance of 38 ± 0.65 F g^{-1} of and 31.7 ± 1.5 F g^{-1} of fibres at a υ of 50 mV s^{-1} in PBS and aqueous 1 M NaCl solutions, respectively (Calculations for specific capacitance per electrode were performed as explained previously in Sect. 2.2.1.11).

Reasonable electroactivity is observed in CV results from coaxial fibres. The curves showed a rectangular shape as observed in Fig. 4.9b which is typically displayed by graphene, implying pure electric double layer capacitive behaviour [66]. This result accompanied by the fast, unexpected colour change of the core material confirms the quick deoxygenation of graphene oxide in the core when exposed to strong alkali solutions at moderate temperatures which was also previously studied [51, 52]. The underlying reduction mechanism in alkaline baths is still unclear. However, it is possible that oxidative debris, which mainly comprises a mixture of complex aromatic structures containing COH-rich functional groups, is stripped from graphene oxide under alkaline conditions [35]. Additionally, the stability of EC performance of coaxial fibres is evidenced by negligible EC changes after 40 cycles in both media as shown in Fig. 4.12a, b.

The specific capacitance, or the ability of the structure to store electrical charge, was also determined for Chit/GO fibres based on CV results in two aqueous media.

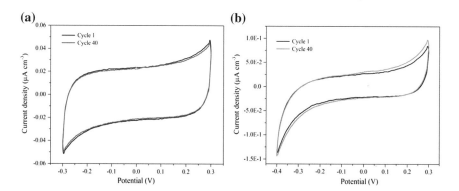

Fig. 4.12 The Cyclic voltammograms of Chit/GO oxide fibres before and after 40 cycles in (**a**) PBS electrolyte and (**b**) 1 M aqueous NaCl solution at 50 mVs^{-1}; potential was scanned between -0.4 V and $+0.3$ (*vs.* Ag/AgCl)

Herein, 1.2 cm of the coaxial fibre was inserted into the electrolyte medium which contains ca. 2.8 mg of the conducting element (rGO). It was found that the coaxial fibre has more specific capacitance in aq. 1 M NaCl solution compared to PBS solution. That might be due to the higher ion exchange rate in 1 M NaCl mainly due to the fact that NaCl medium only contains chloride ions, known as univalent small counterions which have little impact on the mobility of Na^+ ions. Thus, Na^+ ions could be easily inserted into or expelled from the fibres and participate more actively in the redox reaction, while PBS electrolyte contains high amounts of large counterions such as $H_2PO_4^-$ which could retard the ion migration. Chit/GO fibres have specific capacitances of 22.2 ± 1.1 and 27.8 ± 1.5 F g^{-1} of fibres in PBS and aqueous 1 M NaCl solutions, respectively. These performances are significantly lower than those of measured for rGO fibres. This might be due to the harder access of the counterions to the conductive core since it has been covered with the insulating hydrogel. The electroactivity observed in these fibres suggests that they might find application as biosensors, bio-batteries and electrodes as well as stimulation of cells and smart drug release studies as previously shown for fibres from graphene and conducting polymers [24, 67–70].

4.3.7 Raman Spectroscopy Results

Raman spectroscopy was performed on the fibre cross-section as the most direct and non-destructive technique to investigate the significant structural changes including defects, ordered and disordered structure of graphene (C–C bonds are Raman-active bonds) occurring during the chemical reduction. The two distinct peaks at 1341–50 and 1572–79 cm^{-1} assigned to D and G bands and are related to disorder and graphitic order, respectively [71]. It is known that the G band indicates graphite carbon structure (sp^2), whereas the D band is typically an indication of disorder in the Raman spectrum of carbon materials [72, 73]. The normalised Raman spectra for rGO, Chit/GO and reduced Chit/GO fibres is shown in Fig. 4.13.

The normalised ratio of the D-peak to the G-peak (ID/IG) observed in the Raman spectra for Chit/GO, reduced Chit/GO and rGO fibres are presented in Table 4.2.

As indicated, Raman spectra show reduction on the intensity ratio of ID/IG for Chit/GO fibre before and after the chemical reduction from 0.81 to 0.45. There could be seen a small shift of the D peak position comparing coaxial fibres before and

Table 4.2 I_D/I_G as observed in the Raman spectra for Chit/GO, reduced Chit/GO and rGO fibres

Sample Name	I_D	I_G	I_D/I_G
rGO fibre	94.58619	210.39275	0.449
Chit/GO fibre	402.00418	496.02519	0.811
Reduced Chit/GO fibre	321.18929	707.56676	0.454

Fig. 4.13 Raman spectra
from Chit/GO, reduced
Chit/GO and rGO fibres

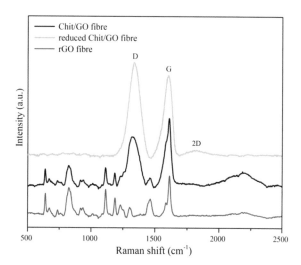

after the reduction process as well. Moreover, according to the Table 4.2, ID/IG after
the reduction process showed similar values to that of rGO fibres. ID/IG is known
to be related to the average size of the sp^2 domains [74], and it is increased with
oxidation, showing an increment in the number of defect sites. Therefore, our results
indicate that the average size of sp^2 domains has changed significantly after chemical
reduction as the intensity ratio of ID/IG is decreased. The reduction of ID/IG might
also be due to the decrease in the size of the newly formed graphene like sp^2 domains
[75]. It is reasonable that thermal and chemical reduction can remove the functional
groups from GO; thus, the exfoliation of GO is inevitable during the reduction and
these factors make the sp^2 domain size of GO to be changed after reduction. Kang
et al. also reported that ID/IG ratio has decreased after thermal reduction of GO
at 200 °C which appeared to be more effective in restoring electrical conductivity
than chemical treatment [76]. Also noticeable is the shoulder in the G band centred
at ~1609 cm^{-1} corresponds to the 2D band, commonly associated to disorder in
graphite. Some of the previous researches reported on the decrease of ID/IG ratio as
a result of deoxygenation of GO. However, 2D band is normally ignored on the studies
of GO and rGO because of the weak intensity [76, 77]. Some other researchers, on
the other hand, such as Ferrari et al. proposed that GO is in an amorphous state and
that a graphite-like state is only recovered after reduction, implying that the ID/IG
ratio cannot be directly compared between the two states [78].

4.3.8 In Vitro Bioactivity Experiments

The proportion of live cells was seen to be very high on both rGO and coaxial fibres,
and on the underlying cells growing on tissue culture plastic, with all cell populations

showing greater than 99% viability (Fig. 4.14a). This was an indication that the fibres did not release any chemicals that caused cell death. The density of cells growing on the fibre surfaces was lower than that observed on the underlying cell bed by nearly a factor of three (Fig. 4.14b), suggested that the surface properties of the fibres were less optimal for cell adhesion and/or proliferation than the glass microscope slide surface, with no difference observed for the cell density on the rGO compared to the coaxial fibres. It is also worth noting that the fibres were quite floating in the media and it was hard to get the cells to stay on them, despite it was tried to weight them down in the dish with cell crowns. However, cells were observed to adhere, and the density of cells on the fibres increased overtime, which together with the high cell viability indicates that the fibres were capable of supporting cell adhesion and proliferation. L-929 morphology on the fibres was typical of L-929 cells, showing good adhesion to both the rGO and coaxial fibres (Fig. 4.14c, d).

In addition, rGO fibres seemed to be quite fragile for being imaged by confocal microscope (as shown in SEM images in Sect. 4.3.2.2 and discussed in mechanical properties in Sect. 4.3.4, too), while Chit/GO fibres seemed to display suitable mechanical properties, also in wet-state (Fig. 4.15).

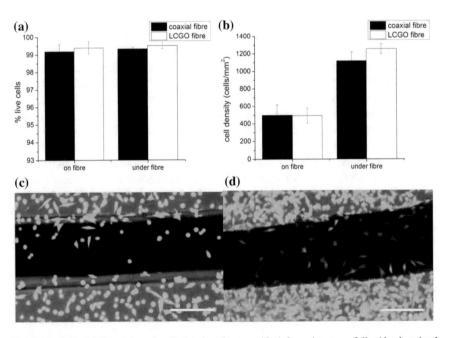

Fig. 4.14 Cell viability (**a**) and cell density (**b**) quantified from images of live/dead stained L-929 cells grown on coaxial (**c**) and rGO (**d**) fibres and cell population growing on the underlying microscope slide. Values in **a** and **b** obtained by image analysis represent the average of at least 20 images (900 × 900 μm each), or at least 50 mm of fibre length, and error bars show one standard deviation of the mean. Scale bars in **c** and **d** represent 200 μm

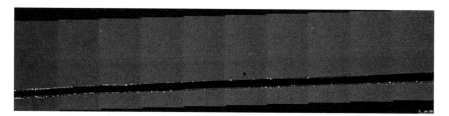

Fig. 4.15 Confocal microscopy image showing cell attachment on the surface of coaxial fibre along its length

(a)   **(b)**

Fig. 4.16 Differentiated neural cell line (PC-12 cells) on **a** coaxial and **b** reduced graphene oxide fibres, showing small degree of differentiation (neurites indicated by arrows)

The adhesion of less-robust PC-12 cells was lower than L-929 cells, however no collagen or adhesion molecules were used (often used in PC-12 studies to promote adhesion) [79, 80]. PC-12 cells were able to undergo some degree of NGF-induced differentiation on both fibres, as shown by the presence of the neurite projections in Fig. 4.16. However, for the long period of differentiation (10 days, compared to 3–5 often used for PC-12 studies) [81, 82], the neurite lengths are shorter than expected, and no extensive network formation was observed. It was expected that the aligned wrinkled morphology observed on the exterior of the fibres (Fig. 4.16a) may align neurite outgrowth, and this was observed somewhat on both the rGO and coaxial fibres, however the effect was not significant, with non-aligned neurites observed as frequently as aligned neurites. Overall, the fibres supported the adhesion and differentiation of the neural model cell line; however the fibres likely require some optimisation for a neural application.

4.4 Conclusion

A one-step wet-spinning process enabled the formation of electroactive coaxial fibres using chitosan and LC GO for the first time. These fibres revealed to have a unique combination of cytocompatiblity, electrical conductivity and acceptable range of mechanical properties through several characterisation methods. Drawn Chit/GO fibres indicated the resultant mechanical strength of ~257 MPa, and the modulus was found to be ~10.7 GPa which is much higher comparing to the rGO fibres. The Redox properties of the bulk material were shown by cyclic voltammetry. Two cell type behaviours have been also investigated on rGO and Chit/GO fibres. For both cell lines, the response was very similar on the coaxial fibres and the rGO fibres. Cells showed to adhere to the fibre surfaces and greater than 99% viability. The results presented here imply that these fibres hold a great promise in applications such as drug delivery, batteries and tissue scaffolds.

References

1. Doucet BM, Lam A, Griffin L (2012) Neuromuscular electrical stimulation for skeletal muscle function. Yale J Biol Med 85:201–215
2. Hsu M, Wei S, Chang Y, Gung C (2011) Effect of neuromuscular electrical muscle stimulation on energy expenditure in healthy adults. Sensors 11:1932–1942. https://doi.org/10.3390/s110 201932
3. Longo U, Loppini M, Berton A, Spiezia F, Maffulli N, Denaro V (2012) Tissue engineered strategies for skeletal muscle injury. Stem Cells Int 2012:175038. https://doi.org/10.1155/201 2/175038
4. Meng S, Rouabhia M, Zhang Z, De D, De F, Laval U (2011) Electrical stimulation in tissue regeneration. In: Applied biomedical engineering. pp 37–62
5. Rupp A, Dornseifer U, Fischer A, Schmahl W, Rodenacker K, Uta J, Gais P, Biemer E, Papadopulos N, Matiasek K (2007) Electrophysiologic assessment of sciatic nerve regeneration in the rat: surrounding limb muscles feature strongly in recordings from the gastrocnemius muscle. J Neurosci Methods 166:266–277. https://doi.org/10.1016/j.jneumeth.2007.07.015
6. Wei Z (2014) Nanoscale tunable reduction ofgraphene oxide for graphene electronics. Science (80-) 1373:1372–1376. https://doi.org/10.1126/science.1188119
7. Chao Y, Chao EY, Inoue N (2003) Biophysical stimulation of bone fracture repair, regeneration and remodelling. Eur Cell Mater 6:72–85
8. Lafayette W (2003) Criteria for the selection of materials for implanted electrodes. Anal Biomed Eng 31:879–890. https://doi.org/10.1114/1.1581292
9. Ben-jacob E, Hanein Y (2008) Carbon nanotube micro-electrodes for neuronal interfacing. J Mater Chem 18:5181–5186. https://doi.org/10.1039/b805878b
10. Wallace G, Moulton S, Kapsa R, Higgins M (2012) Key elements of a medical bionic device. In: Organic bionics. p 240
11. Ateh D, Navsaria H, Vadgama P (2006) Polypyrrole-based conducting polymers and interactions with biological tissues. J R Soc Interface 3:741–752. https://doi.org/10.1098/rsif.2006.0 141
12. Min Y, Yang Y, Poojari Y, Liu Y, Wu J, Hansford DJ, Epstein AJ (2013) Sulfonated polyaniline-based organic electrodes for controlled electrical stimulation of human osteosarcoma cells. Biomacromol 14:1727–1731
13. Quigley BAF, Razal JM, Thompson BC, Moulton SE, Kita M, Kennedy EL, Clark GM, Wallace GG, Kapsa RMI (2009) A conducting-polymer platform with biodegradable fibers for stimulation and guidance of axonal growth. Adv Mater 21:1–5. https://doi.org/10.1002/adma. 200901165
14. Heo C, Yoo J, Lee S, Jo A, Jung S, Yoo H, Hee Y, Suh M (2011) The control of neural cell-to-cell interactions through non-contact electrical field stimulation using graphene electrodes. Biomaterials 32:19–27. https://doi.org/10.1016/j.biomaterials.2010.08.095

15. Park D, Schendel AA, Mikael S, Brodnick SK, Richner TJ, Ness JP, Hayat MR, Atry F, Frye ST, Pashaie R, Thongpang S, Ma Z, Williams JC (2014) Graphene-based carbon-layered electrode array technology for neural imaging and optogenetic applications. Nat Commun 5:1–11. https://doi.org/10.1038/ncomms6258

16. Sherrell P, Thompson B, Wassei J, Gelmi A, Higgins M, Kaner R, Wallace G (2014) Maintaining cytocompatibility of biopolymers through a graphene layer for electrical stimulation of nerve cells. Adv Funct Mater 24:769–776. https://doi.org/10.1002/adfm.201301760

17. Han D, Boyce ST, Steckl AJ (2008) Versatile core-sheath biofibers using coaxial electrospinning. Mater Res Soc Symp Proc 1094:33–38

18. Sun L, Wang S, Zhang Z, Wang X, Zhang Q (2009) Biological evaluation of collagen—chitosan scaffolds for dermis tissue engineering. Biomed Mater 4:2–8. https://doi.org/10.1088/1748-6041/4/5/055008

19. Yarin A (2011) Coaxial electrospinning and emulsion electrospinning of core–shell fibers. Polym Adv Technol 22:310–317. https://doi.org/10.1002/pat.1781

20. Chwojnowski A, Wojciechowski C (2009) Polysulphone and polyethersulphone hollow fiber membranes with developed inner surface as material for bio-medical applications. Biocybern Biomed Eng 29:47–59

21. Esrafilzadeh D, Razal J, Moulton S, Stewart E, Wallace G (2013) Multifunctional conducting fibres with electrically controlled release of ciprofloxacin. J Control Release 169:313–320

22. Granero A, Razal J, Wallace G, Panhuis M (2010) Conducting gel-fibres based on carrageenan, chitosan and carbon nanotubes.pdf. J Mater Chem 20:7953–7956. https://doi.org/10.1039/c0jm00985g

23. Nohemi R, Araiza R, Rochas C, David L, Domard A (2008) Interrupted wet-spinning process for chitosan hollow fiber elaboration. Macromol Symp 266:1–5. https://doi.org/10.1002/masy.200850601

24. Kou L, Huang T, Zheng B, Han Y, Zhao X, Gopalsamy K, Sun H, Gao C (2014) Coaxial wet-spun yarn supercapacitors for high-energy density and safe wearable electronics. Nat Commun 5:3754. https://doi.org/10.1038/ncomms4754

25. Gupta V (1997) Solution-spinning processes. In: Gupta VB, Kothari VK (eds) Manufactured fibre technology. Chapman & Hall, London, pp 124–138

26. Agboh O, Qin Y (1998) Chitin and chitosan fibers. Polym Adv Technol 8:355–365

27. Khor E, Yong L (2003) Implantable applications of chitin and chitosan. Biomaterials 24:2339–2349. https://doi.org/10.1016/S0142-9612(03)00026-7

28. Kumar M (1999) Chitin and chitosan fibres: a review. Bull Mater Sci 22:905–915

29. Sonia T, Sharma C (2011) Chitosan and its derivatives for drug delivery perspective. Adv Polym Sci 243:23–54. https://doi.org/10.1007/12_2011_117

30. Dash M, Chiellini F, Ottenbrite RM, Chiellini E (2011) Chitosan—a versatile semi-synthetic polymer in biomedical applications. Prog Polym Sci 36:981–1014. https://doi.org/10.1016/j.progpolymsci.2011.02.001

31. Dong Z, Jiang C, Cheng H, Zhao Y, Shi G, Jiang L, Qu L (2012) Facile fabrication of light, flexible and multifunctional graphene fibers. Adv Mater 24:1856–1861. https://doi.org/10.1002/adma.201200170

32. Zhu Y, Murali S, Cai W, Li X, Suk J, Potts J, Ruoff R (2010) Graphene and graphene oxide: synthesis, properties, and applications. Adv Funct Mater 22:3906–3924. https://doi.org/10.1002/adma.201001068

33. Zhang W, He W, Jing X (2010) Preparation of a stable graphene dispersion with high concentration by ultrasound. J Phys Chem B 114:10368–10373

34. Cong H-P, Ren X-C, Wang P, Yu S-H (2012) Wet-spinning assembly of continuous, neat, and macroscopic graphene fibers. Sci Rep 2:613. https://doi.org/10.1038/srep00613

35. Jalili R, Aboutalebi H, Esrafilzadeh D, Shepherd R, Chen J, Aminorroaya-yamini S, Konstantinov K, Minett A, Razal J, Wallace G (2013) Scalable one-step wet-spinning of graphene fibers and yarns from liquid crystalline dispersions of graphene oxide: towards multifunctional textiles. Adv Funct Mater 23:5345–5354. https://doi.org/10.1002/adfm.201300765

36. Iqbal K, Khan A, Ali M, Khattak K (2004) Biological significance of ascorbic acid (Vitamin C) in human health—a review. Pak J Nutr 3:5–13
37. Han D, Yan L, Chen W, Li W (2011) Preparation of chitosan-graphene oxide composite film with enhanced mechanical strength.pdf. Carbohydr Polym 83:653–658
38. Costa-Pinto A, Reis R, Neves N (2011) Scaffolds based bone tissue engineering: the role of chitosan. Tissue Eng B 17:331–347. https://doi.org/10.1089/ten.teb.2010.0704
39. Suh J, Matthew H (2000) Application of chitosan-based polysaccharide biomaterials in cartilage tissue engineering: a review. Biomaterials 21:2589–2598
40. Zhu X, Liu Q, Zhu X, Li C, Xu M, Liang Y (2012) Reduction of graphene oxide via ascorbic acid and its application for simultaneous detection of dopamine and ascorbic acid. Int J Electrochem Sci 7:5172–5184
41. Oyane A, Kim H-M, Furuya T, Kokubo T, Miyazaki T, Nakamura T (2003) Preparation and assessment of revised simulated body fluids. J Biomed Mater Res A 65:188–195. https://doi.org/10.1002/jbm.a.10482
42. Reaction R (2013) Covalently bonded chitosan on graphene oxide via redox reaction. Materials (Basel) 6:911–926. https://doi.org/10.3390/ma6030911
43. Yu D, Branford-White K, Chatterton N, Zhu L, Huang L, Wang B (2011) A modified coaxial electrospinning for preparing fibers from a high concentration polymer solution. Express Polym Lett 5:732–741. https://doi.org/10.3144/expresspolymlett.2011.71
44. El-Tahlawy K, Hudson S (2006) Chitosan: aspects of fiber spinnability. J Appl Polym Sci 100:1162–1168. https://doi.org/10.1002/app.23201
45. Hirano S, Bash E (2001) wet-spinning and applications of functional fibers based on chitin and chitosan. Macromol Symp 168:21–30. https://doi.org/10.1017/CBO9781107415324.004
46. Jayakumar R, Prabaharan M, Kumar P, Nair S, Tamura H (2011) Biomaterials based on chitin and chitosan in wound dressing applications. Biotechnol Adv 29:322–337. https://doi.org/10.1016/j.biotechadv.2011.01.005
47. Shao X, Hunter CJC (2007) Developing an alginate/chitosan hybrid fiber scaffold for annulus fibrosus cells. J Biomed Mater Res A 82:702–710. https://doi.org/10.1002/jbm.a
48. Mirabedini A, Foroughi J, Romeo T, Wallace GGGG (2015) Development and characterization of novel hybrid hydrogel fibers. Macromol Mater Eng 300:1217–1225. https://doi.org/10.1002/mame.201500152
49. Yu J, Fridrikh S, Rutledge G (2004) Production of submicrometer diameter fibers by two-fluid electrospinning. Adv Mater 78:1562–1566. https://doi.org/10.1002/adma.200306644
50. Elahi F, Lu W, Guoping G, Khan F (2013) Core-shell fibers for biomedical applications—a review. J Bioeng Biomed Sci 3:1–14. https://doi.org/10.4172/2155-9538.1000121
51. Fan X, Peng W, Li Y, Li X, Wang S, Zhang G, Zhang F (2008) Deoxygenation of exfoliated graphite oxide under alkaline conditions: a green route to graphene preparation. Adv Mater 20:4490–4493. https://doi.org/10.1002/adma.200801306
52. Rourke J, Pandey P, Moore J, Bates M, Kinloch I, Young R, Wilson NR (2011) The real graphene oxide revealed: stripping the oxidative debris from the graphene-like sheets. Angew Chem Int Ed 50:3173–3177. https://doi.org/10.1002/anie.201007520
53. Kaczmarek D (1998) Backscattered electrons topographic mode problems in the scanning electron microscope. Opt Appl 12:161–169
54. Xu Z, Gao C (2011) Graphene chiral liquid crystals and macroscopic assembled fibres. Nat Commun 2:571–579. https://doi.org/10.1038/ncomms1583
55. Xu Z, Sun H, Zhao X, Gao C (2013) Ultrastrong fibers assembled from giant graphene oxide sheets. Adv Mater 25:188–193. https://doi.org/10.1002/adma.201203448
56. Whitby RLD (2014) Chemical control of graphene architecture: tailoring shape and properties. ACS Nano 8:9733–9754
57. Shao J, Lv W, Yang Q (2014) Self-assembly of graphene oxide at interfaces. Adv Mater 5586–5612. https://doi.org/10.1002/adma.201400267
58. Velasco-santos C, Almendarez-camarillo A, Faculty C, Tollocan P, De Queretaro S, Madero C, Del Charro A (2014) All green composites from fully renewable biopolymers: chitosan-starch reinforced with keratin from feathers. Polymers (Basel) 6:686–705. https://doi.org/10.3390/polym6030686

59. Salehizadeh H, Hekmatian E, Sadeghi M, Kennedy K (2012) Synthesis and characterization of core-shell Fe_3O_4-gold-chitosan nanostructure. J Nanobiotechnol 10:2–8
60. Zuo P, Feng H, Xu Z, Zhang L, Zhang Y, Xia W, Zhang W (2013) Fabrication of biocompatible and mechanically reinforced graphene oxide-chitosan nanocomposite films. Chem Cent J 7:1–11
61. (1964) Carbon dioxide. NIST database
62. Rosales-Leal J, Rodríguez-Valverde M, Mazzaglia G, Ramón-Torregrosa P, Díaz-Rodríguez L, García-Martínez O, Vallecillo-Capilla M, Ruiz C, Cabrerizo-Vílchez M (2010) Effect of roughness, wettability and morphology of engineered titanium surfaces on osteoblast-like cell adhesion. Colloid Surf A Physicochem Eng Asp 365:222–229. https://doi.org/10.1016/j.colsurfa.2009.12.017
63. Zhou M, Zhai Y, Dong S (2009) Electrochemical sensing and biosensing platform based on chemically reduced graphene oxide. Anal Chem 81:5603–5613
64. Yang X, Zhu J, Qiu L, Li D (2011) Bioinspired effective prevention of restacking in multilayered graphene films: towards the next generation of high-performance supercapacitors. Adv Mater 23:2833–2838. https://doi.org/10.1002/adma.201100261
65. Ultracapacitors G, Stoller MD, Park S, Zhu Y, An J, Ruoff RS, Stoller MD, Park S, Zhu Y, An J, Ruoff RS (2008) Graphene-based ultracapacitors. Nano Lett 8:3498–3502. https://doi.org/1 0.1021/nl802558y
66. Xu B, Yue S, Sui Z, Zhang X, Hou S, Cao G, Yang Y (2011) What is the choice for supercapacitors: graphene or graphene oxide? Energy Environ Sci 4:2826. https://doi.org/10.1039/c1 ee01198g
67. Ahuja T, Kumar D (2009) Recent progress in the development of nano-structured conducting polymers/nanocomposites for sensor applications. Sens Actuators, B 136:275–286. https://doi.org/10.1016/j.snb.2008.09.014
68. Foroughi J, Spinks G, Antiohos D, Mirabedini A, Gambhir S, Wallace G, Ghorbani S, Peleckis G, Kozlov M, Lima M, Baughman R (2014) Highly conductive carbon nanotube-graphene hybrid yarn. Adv Funct Mater 24:5859–5865. https://doi.org/10.1002/adfm.201401412
69. Foroughi J, Spinks G, Wallace G (2011) Chemical high strain electromechanical actuators based on electrodeposited polypyrrole doped with di-(2-ethylhexyl) sulfosuccinate. Sens Actuators, B 155:278–284. https://doi.org/10.1016/j.snb.2010.12.035
70. Gupta N, Sharma S, Mir I, Kumar D (2006) Advances in sensors based on conducting polymers. J Sci Ind Res 65:549–557
71. Torres D, Pinilla JL, Moliner R, Suelves I (2014) On the oxidation degree of few-layer graphene oxide sheets obtained from chemically oxidized multiwall carbon nanotubes. Carbon 1:2–10
72. Fan J, Michalik JM, Casado L, Roddaro S, Ibarra MR, De Teresa JM (2011) Investigation of the influence on graphene by using electron-beam and photo-lithography. Solid State Commun 151:1574–1578. https://doi.org/10.1016/j.ssc.2011.07.028
73. Wang SJ, Geng Y, Zheng Q, Kim J (2010) Fabrication of highly conducting and transparent graphene films. Carbon N Y 8:1815–1823. https://doi.org/10.1016/j.carbon.2010.01.027
74. Tuinstra F, Koenig JL (2011) Raman spectrum of graphite. J Chem Phys 1126:1126–1130. https://doi.org/10.1063/1.1674108
75. Sobon G, Sotor J, Jagiello J, Kozinski R, Zdrojek M, Holdynski M, Paletko P, Boguslawski J, Lipinska L, Abramski KM (2012) Graphene oxide vs. reduced graphene oxide as saturable absorbers for Er-doped passively mode-locked fiber laser. Opt Express 20:19463–19473
76. Kang H, Kulkarni A, Stankovich S, Ruoff RS, Baik S (2009) Restoring electrical conductivity of dielectrophoretically assembled graphite oxide sheets by thermal and chemical reduction techniques. Carbon N Y 7:7–12. https://doi.org/10.1016/j.carbon.2009.01.049
77. Stankovich S, Dikin DA, Piner RD, Kohlhaas KA, Kleinhammes A, Jia Y, Wu Y (2007) Synthesis of graphene-based nanosheets via chemical reduction of exfoliated graphite oxide. Carbon N Y 45:1558–1565. https://doi.org/10.1016/j.carbon.2007.02.034
78. Ferrari AC, Robertson J, Trans P, Lond RS (2004) Raman spectroscopy of amorphous, nanostructured, diamond–like carbon, and nanodiamond. Phil Trans R Soc Lond A 362:2477–2512. https://doi.org/10.1098/rsta.2004.1452

79. Attiah D, Kopher R, Desai T (2003) Characterization of PC12 cell proliferation and differentiation-stimulated by ECM adhesion proteins and neurotrophic factors. J Mater Sci Mater Med 4:1005–1009
80. Cooke M, Phillips S, Shah D, Athey D, Lakely J, Przyborski S (2008) Enhanced cell attachment using a novel cell culture surface presenting functional domains from extracellular matrix proteins. Cytotechnology 56:71–79. https://doi.org/10.1007/s10616-007-9119-7
81. Liu JW, Montero M, Bu L, De Leon M (2015) Epidermal fatty acid-binding protein protects nerve growth factor-differentiated PC12 cells from lipotoxic injury. J Neurochem 132:85–98. https://doi.org/10.1111/jnc.12934
82. Park K, Park H, Shin K, Choi H, Kai M, Lee M (2012) Modulation of PC12 cell viability by forskolin-induced cyclic AMP levels through ERK and JNK pathways: an implication for L-DOPA-induced cytotoxicity in nigrostriatal dopamine neurons. Toxicol Sci 128:247–257. https://doi.org/10.1093/toxsci/kfs139

Chapter 5
Development of One-Dimensional Triaxial Fibres as Potential Bio-battery Structures

5.1 Introduction

Technological improvements in the development of smaller and lighter smart multifunctional structures have enabled the fabrication of many miniaturized portable electronic devices in recent years. The integration of these organo-electronic components into common textile structures could facilitate free and easy access while allowing a number of smart functionalities such as signalling, sensing, actuating, energy storage or information processing [1]. Fabrics provide a very practical basis to work with since they can be easily shaped into the human body form as well as providing easy access to the electronic equipment embedded within [2]. This ease-of-use solution, known as electronic textiles (E-textiles) or smart textiles, was first introduced in the early 1990s [3]. The significance of this research is the interdisciplinary approach that it requires; covering textile design, chemistry, materials science and computer science. Wearable batteries or power supplies are one of the more recent applications which could clearly benefit from this transformational trend. Implantable medical devices such as pacemakers, microstimulators and drug delivery microchips could be a potentially lucrative target market for these power sources since they need to meet strict clinical and dimensional constraints [4]. Unlike conventional batteries such as lithium cell, nuclear cell or bio-fuel cell types, which are typically made of rigid materials; so-called one-dimensional (1D) wearable batteries offer flexibility as well as low device weight. Combining solution-processable conductive materials such as conducting polymers and graphene with fabrication methods including printing and spinning has enabled the preparation of mechanically flexible, inexpensive and lightweight battery systems that are customizable [5–7]. A numbers of approaches have been employed to develop non-bulky batteries. Previous attempts have focussed on the fabrication of sandwich-structured batteries [8] or development of flexible paper-like films as electrodes for use in lithium-ion batteries [9–11]. In addition to this, novel fabrication methods have been applied to incorporate the battery components all into one single continuous unit, eliminating the need for packaging [4].

© Springer Nature Switzerland AG 2018
A. Mirabedini, *Developing Novel Spinning Methods to Fabricate Continuous Multifunctional Fibres for Bioapplications*, Springer Theses,
https://doi.org/10.1007/978-3-319-95378-6_5

Poly(3,4-ethylenedioxythiophene) polystyrene sulfonate (PEDOT:PSS) is the polymer of choice when intrinsic properties such as ease of processibility, thermal and atmospheric stability, good optical transmission and high conductivity are considered [12–14]. Among the most significant of these properties are its water solubility and ease of use in wet-spinning [15, 16]. This has enabled the use of PEDOT:PSS in many devices such as biosensors [17, 18], OLED displays [19] and photovoltaics [20, 21]. Some research groups have also attempted to fabricate wet-spun PEDOT:PSS fibres for a wide range of use in applications such as chemical sensors, energy storage electrodes and actuators [12, 16, 22]. Okuzaki and Ishihara reported the first preparation of PEDOT:PSS microfibres *via* wet-spinning showed electrical conductivities of ~195 S cm^{-1} [22]. Shortly thereafter, they fabricated highly conducting PEDOT:PSS microfibres with 5 mm diameter and a 2–6 fold increase in electrical conductivity up to 467 S cm^{-1} *via* wet-spinning followed by ethylene glycol posttreatment [16]. Later on, Jalili et al. simplified the method to a one-step process and prepared continuous microfibres by employing a wet-spinning formulation consisting of an aqueous blend of PEDOT:PSS and poly (ethylene glycol) eliminating the need for post-spinning treatment with ethylene glycol and obtained electrical conductivities of up to 264 S cm^{-1} [12]. Recently, highly stretchable PU/PEDOT:PSS fibres have been prepared which have shown suitable properties that allow knitting of various textile structures for use in strain sensing applications [2].

In this study, chitosan and alginate were chosen as the core material within fibres mostly due to their high ionic conductivities (of the order of 10^{-3} to 10^{-4} S cm^{-1}) when wet. Both biopolymers have aroused a great deal of interest for their potential in industrial, clinical (approved by FDA) and biological applications [23–28].

PPy is also of particular interest since it is simple to prepare, has reasonable stability in the oxidised state, and has high electrical conductivity and strong adhesion to different substrates. Not surprisingly, PPy has already found applications in a wide variety of areas such as rechargeable lithium batteries [29, 30], tissue engineering applications [31, 32] as well as drug delivery systems [33, 34] and organic compound detection [35, 36]. However, PPy prepared by conventional methods is insoluble in most organic solvents [37, 38] owing to the presence of strong interchain interactions and a rigid structure. The poor processability of PPy has motivated research into developing methods to render PPy processable by direct chemical vapour deposition (CVD) [39–42] or electrodeposition [43, 44] of pyrrole monomer onto various substrates and matrices [40]. Direct deposition techniques offer the advantage of applying the conducting material directly onto an electrode surface eliminating issues regarding adhesion of battery layers. CVD, which was employed in this research, is a well-established technique in which a monomer is introduced onto an oxidant-coated substrate in vapour form (discussed in detail previously in Sect. 1.3.1.2.4). The CVD method provides a level of control over conducting polymer film thickness, uniformity and density resulting in higher conductivities [45].

Wet-spinning techniques have been at the centre of attention for years for fabrication of different kinds of fibrillar electroactive electrodes. However, several key factors must be taken into account in designing a high performance fibre-based bio-battery. Practical considerations for the actual fabrication of a device involve selection of the right materials with suitable viscosities and surface charges, proper coagulation processes, appropriate rates of both spinning and drawing ratios for coaxial spinning, as well as accurate experimental conditions for CVD of the selected polymer. Battery performance also needs to be monitored and regulated carefully. High ionic conductivity and well-controlled microporosity are crucial factors in the selection of a separator to provide optimal electrolyte uptake and rapid transfer of ions between two terminals. There would also be a necessity for both electrodes to show high electrical conductivity, high stability and large potential windows.

Here, with the final aim of producing compact thin fibres for a potential battery structure, a facile methodology using a coaxial wet-spinning approach followed by CVD has been established. Using this approach, unlimited lengths of fibre batteries are achievable, where all the battery elements are integrated into one unified structure. Once produced, these fabricated fibre batteries may be easily incorporated into a textile form using knitting or braiding techniques. This study initially explores the conditions necessary to achieve optimal properties in core-sheath fibres produced *via* a one-step spinning process along with their characterisation in terms of morphological, mechanical and electrochemical properties. Secondly, pyrrole was then incorporated as a third layer onto the coaxial fibres to complete the triaxial fibre battery structure. Finally, performances of the hydrogel electrolytes, as well as the EC properties of the conductor, are investigated.

5.2 Experimental

5.2.1 Materials

Iron (III) p-toluenesulfonate hexahydrate (FepTS) (Clevios C-B 40 V2) was obtained from *Heraeus Technology Company* (Germany) and used as an oxidant as supplied. PEDOT:PSS pellets were purchased from *Agfa* (Orgacon dry™, Lot A6 0000 AC), and PEG with a molecular weight of 2000 g mol^{-1} was purchased from *Fluka Analytical*. Sterile filters of 5 μm were supplied from *EASYstrainer™*. Sodium and lithium chloride salts, NaCl and LiCl respectively, were also sourced from *Sigma-Aldrich Co. LLC* (Australia).

5.2.2 Dispersion Preparation

For the purpose of coaxial spinning, chitosan and alginate solutions were prepared such that the final concentrations of both were 3% w v^{-1} as described previously in Sect. 2.1.2. Aqueous dispersions of PEDOT:PSS (with and without the addition of PEG) were subsequently prepared as explained in Sect. 2.1.4.

5.2.3 Coaxial Wet-Spinning of Chit-PEDOT and Alg-PEDOT

Core-sheath fibres were successfully spun using a coaxial spinneret as explained in Sect. 2.2.1.2. PEDOT:PSS dispersions (with and without PEG) were injected through port B and extruded through the centre outlet nozzle into the appropriate coagulation bath. Simultaneously, either alginate or chitosan were extruded as the sheath of the fibre, providing an outer casing for the core, by injection through port A which facilitates extrusion through the outer segment of the spinneret nozzle. Coaxial fibres of chitosan-PEDOT:PSS (with or without PEG) and alginate-PEDOT:PSS (+PEG) (abbreviated respectively as Chit-PEDOT and Alg-PEDOT, for ease of use) were spun into a bath of 1 M aqueous NaOH (Ethanol/H$_2$O: 1/5) for Chit-PEDOT and 2% w v^{-1} CaCl$_2$ for Alg-PEDOT, respectively. The applied injection rate used for PEDOT:PSS dispersion was 15 and 28 mL h^{-1} for either chitosan or alginate solution in order to provide sufficient time to cover the core material. It is worth noting that injection rates of 20 and 25 mL h^{-1} were tested initially for both alginate and chitosan solutions while the injection rate of core material was kept at 15 mL h^{-1} (the lowest rate at which a continuous filling of the core could be obtained). However, at this rate, it seemed that the sheath formed was not thick enough to hold the core material in place. 28 mL h^{-1} was determined to be the optimum rate for the sheath component injection resulting in the thinnest sheath that was still capable of supporting the core material structure. Once the sheath components were coagulated, the fibres were then transferred into an isopropanol bath for post-treatment. After that, they were soaked in a graded series consisting of 80/20, 60/40, 40/60, 20/80, and 0/100 ISP/MilliQ water mixtures gradually over 24 h as a washing step. This post-treatment procedure is a useful step in helping to lock the core component in place.

In an acidic medium, chitosan amino groups are positively charged and can thus react with the "free" negatively charged polystyrene sulphonate acid (PSS) groups of PEDOT:PSS, according to the polyionic complexation coagulation strategy, [46] to create hydrogen bonding at the interface between the core and the sheath in a coaxial fibre [47]. The electrostatic absorption between PEDOT:PSS and chitosan is shown in Fig. 5.1.

In contrast, sodium alginate [being R–COO$^-$Na$^+$] is negatively charged as noted earlier [48]. This will lead to repulsive effects when carboxyl groups of alginate are placed in the vicinity of the PSS groups of PEDOT: PSS. This prediction was confirmed in later parts of the study by looking at LV-SEM images.

5.2.4 Polymerisation of Pyrrole

The CVD method was used to coat the coaxial fibre with PPy and create triaxial electroactive fibres. Coaxial fibres of Chit/PEDOT were then soaked with oxidant solutions until the oxidizing agent could be fully absorbed into the fibre surface. After oxidant impregnation, the fibres were then dried at 80 °C resulting in a colour

Fig. 5.1 The electrostatic absorption between PEDOT: PSS and chitosan

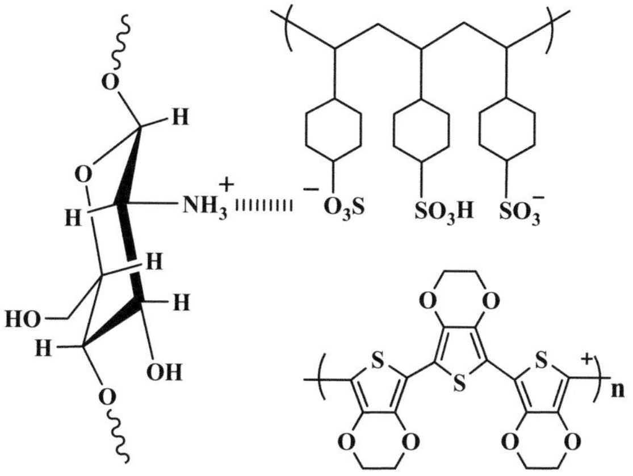

change from yellow to orange-yellow which is an indication of solvent evaporation and activation of the oxidant. The oxidant-enriched fibres were then placed in a closed reactor chamber and exposed to pyrrole. The pyrrole monomer vapour was generated by heating liquid distilled pyrrole at 70 °C while the monomer container was placed in the vicinity of the fibres. Exposure of oxidant-enriched fibres to the atmosphere containing the pyrrole monomer, the spontaneously initiated the polymerization reaction. There was a visible, rapid colour change from orange to black as the fibres were coated with a dark layer of PPy polymer. Finally, the coated fibres were rinsed several times with ethanol in order to remove any unreacted monomer, by-products and Fe (III) salts. The same protocol was applied to the production of triaxial fibres of PPy-Alg-PEDOT (+PEG) fibres without success. It appears that the polymerisation of pyrrole could not be initiated on the surface of the alginate as observed previously for chitosan. The reasons for this observation are investigated in more detail in the Sect. 5.3.1.3.2.

5.2.5 Fourier Transform Infrared

FTIR spectra of the sodium alginate and alginate treated with FepTS were recorded to evaluate the possible reactions between alginate and the oxidant. This experiment would clearly show why pyrrole polymerisation will not occur on the surface of alginate. FTIR spectroscopy was performed in a KBr pellet on a Shimadzu FTIR Prestige-21 spectrometer, in the 700–4000 cm^{-1} range with 4 cm^{-1} resolution.

5.2.6   Analysis

5.2.6.1   Characterisations of Spinning Solutions

Rheological Measurement

There are upper and lower practical limits which need to be considered in terms of suitable polymer concentrations for wet-spinning applications depending on the type of polymer as discussed in Sect. 3.3.1 [49]. In addition to this, the viscosity is regarded as the primary criterion for the selection of suitable concentrations of materials for the purpose of coaxial spinning. Thus, an understanding of the rheological properties of spinning solutions is essential to determine the optimum conditions required for the spinning process. Changes in viscosity have been recorded versus shear rate. The rheological properties of chitosan (3% w v^{-1}), alginate (3% w v^{-1}) (with and without NaCl) and PEDOT:PSS (3% w v^{-1}) (with and without PEG) solutions were examined in flow mode (cone and plate method) by Rheometer-AR G2 (TA Instruments, USA). Changes in shear viscosity were also investigated for each sample (in triplicate) at room temperature (~25 °C) for shear rates between 0.1 and 300 s^{-1}.

Impedance Behaviour of Chitosan and Alginate Hydrogels

The electrical impedance behaviour of hydrogel samples (width = 5 mm, height = 6 mm) was measured using a custom-built impedance set-up, described elsewhere [50]. In brief, an oscilloscope (Agilent U2701A) and waveform generator (Agilent U2761A) were used to apply a range of frequencies at a peak-to-peak a.c. voltage of 0.8 V, while measuring the voltage drop across a known resistor (10 kΩ) and the unknown hydrogel sample. The current was calculated across the known resistor and this was then used to calculate impedance behaviour of the hydrogel samples as a function of frequency. Hydrogel samples tested varied in length (l) between 0.5 and 2.5 cm and were contacted at each end with reticulated vitreous carbon (RVC, ERG Aerospace, 20 pores per inch), as shown in Fig. 5.2.

The RVC acted as a medium for contact between the hard electrodes and soft gels. Electrical conductivity (σ) was calculated according to Eq. 5.1 as below:

$$Z_I = \frac{l}{\sigma A_C} + R_C \tag{5.1}$$

where Z_I is the frequency-independent impedance, R_C the contact resistance and A_C is the cross-sectional area of the sample.

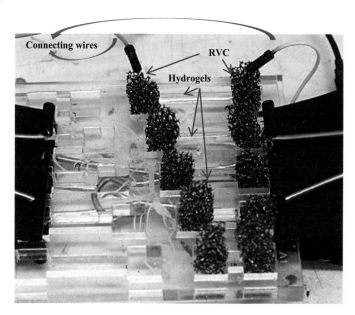

Fig. 5.2 Hydrogel sample holder containing 5 channels varying in length between RVC electrodes from 0.5 to 2.5 cm (height, 6 mm; channel width, 5 mm)

5.2.6.2 Characterisation of Fibres

5.2.6.2.1. Scanning Electron Microscopy/Energy-Dispersive X-ray Spectroscopy (SEM/EDS)

The morphological properties of the dried fibres were observed using a JEOL JSM-7500 FESEM, as described in Sect. 2.2.1.6. Compositional analysis of the cross-sections of Chit/PEDOT(+PEG) and Alg/PEDOT(+PEG) coaxial fibres was then carried out using a Bruker X-Flash 4010 EDS system, fitted with a 10 mm^2 SD detector (127 eV resolution) equipped with Esprit 1.9 microanalysis software. This allowed us a better understanding of where the core and sheath components were locally dispersed by creating an elemental distribution map of the cross-section surface.

5.2.6.2.2. Cyclic Voltammetry (CV)

Electrochemical properties of as-prepared fibres were investigated *via* the use of cyclic voltammetry as explained previously in Sect. 2.2.1.11. The experiment was performed on Chit/PEDOT (+PEG) and Alg/PEDOT (+PEG) fibres in aqueous and organic media of phosphate buffered saline (PBS) and 0.1 M tetrabutylammonium bromide (TBABF$_4$), respectively. A three-electrode cell was used to study the electrochemical behaviour of the coaxial fibres. A steel wire with a diameter of ~25 μm was inserted into the fibre core while spinning to make the electrical connection to the core of the fibres. CV measurements were carried out using an E-Corder 401 interface and a potentiostat (EDAQ) in deoxygenated PBS and 0.1 M TBABF$_4$ in acetonitrile using a Ag/Ag$^+$ reference electrode and a Pt mesh auxiliary electrode. Fibres showed sufficient stability in both solutions. The applied υ was 10 mV s^{-1} and 50 cycles were performed.

5.2.6.2.3. In Vitro Experiments with Living Cells

The cytotoxicity of the coaxial fibres and their cytocompatiblity for cell proliferation were determined with the use of a primary myoblast muscle cell line (Rosa, kindly donated by Prof. Robert Kapsa, St. Vincent's Hospital, Victoria, Australia) by Rodrigo Lozano.

Cell adhesion and proliferation on both Alg-PEDOT (+PEG) and Chit-PEDOT (+PEG) coaxial fibres were evaluated without the addition of any extra-cellular matrices to the fibres. Prior to the use of fibres in the cell culture experiments, 30 mm lengths of fibres were fixed onto microscope slides with 4-chamber wells (Lab Tek$^®$II, Thermo Fisher Scientific) glued on top and allowed to dry overnight. The samples then underwent a sterilization process consisting of two washes in sterile 70% ethanol (each for 30 min) and then four washes (each for 30 min) in sterile phosphate buffered saline (PBS, *Sigma-Aldrich Co.*). Finally, they were kept in cell culture media overnight to remove all excess acid.

Myoblast cells were seeded at 5×10^5 cells/cm^2 in a proliferation medium containing Ham's F-10 medium (*Sigma-Aldrich Co.*), supplemented with 2.5 ng mL^{-1} bFGF (Peprotech) and 20% fetal bovine serum (FBS, Invitrogen supplied by Life Technologies) and 1% penicillin/streptomycin (P/S, Life Technologies). This medium was changed to a differentiation medium after 24 h in culture (50–50 mixture of Dulbecco's modified Eagle's medium (DMEM, Life Technologies) and F-12 medium (DMEM/F12, Life Technologies) supplemented with 3% FBS and 1% P/S). The cells were left and allowed to differentiate for 4 days in the media. Cell morphology was assessed after 5 days in three independent experiments containing four samples each. Myoblast cells were stained by addition of calcein AM (Invitrogen) solution at a final concentration of 5 μM (1:200 dilution) and incubated at 37 °C for 10 min. Image analysis was performed using a Leica TSC SP5 II microscope.

5.3 Results and Discussions

5.3.1 Characterisation of Spinning Solutions

5.3.1.1 Rheological Properties

5.3.1.1.1. Concentrations of Spinning Solutions

Broad ranges varying from 2–15% (w v^{-1}) have been reported as practical, spinnable concentrations for chitosan [51, 52]. Viscosity changes of both hydrogel precursors upon increasing shear rate have been discussed in detail as described in detail in Sect. 3.3.1. A concentration of 3% (w v^{-1}) has been selected here as an average value for both gels for ease of spinnability. PEDOT:PSS concentrations of less than 1.5% (w v^{-1}) showed insufficient viscosity to promote fibre production while concentrations higher than 3% (w v^{-1}) were too viscous. A concentration of 2.5% (w v^{-1}) PEDOT:PSS dispersion has been suggested to provide the most "spinnable" formulation, having been demonstrated by the production of continuous fibres over several meters in length in earlier studies [12]. Isopropanol was also the recommended choice for the coagulation bath solvent to be used in the production of these continuous fibre lengths [53]. In spite of this, after observation of coaxial fibre cross-sections under the scanning electron microscope, it was seen that the PEDOT:PSS core section had an inconsistent texture. In order to achieve uniform inner structure, as well as improving its conductivity, we investigated the effect of adding PEG into the PEDOT:PSS spinning formulation (fibres abbreviated "Hydrogel-PEDOT (+PEG)" in the succeeding sections). We found that by addition of 10% (w v^{-1}) PEG, the spinnability of the formulation was unaffected, thereby affording a one-step fibre production method. In addition to this, a uniform core texture could now be obtained routinely enclosed within the outer sheath component.

5.3.1.1.2. Viscosity of Spinning Solutions

Viscosity is another key factor determining the right selection of spinning solutions for wet-spinning. The viscosity of the sheath spinning solution is particularly critical in coaxial spinning as it is needed to provide a continuous protective coating over the inner core material. Additionally, the core component material must also possess a certain minimum viscosity to allow continuous spinnability without break-up as discussed previously. For coaxial spinning, matching the viscosities of the two components is also an essential consideration. Figure 5.3 shows changes in viscosity versus shear rate at different shear rates between 0.1 and 500 s^{-1} for aqueous solutions of chitosan at 3% (w v^{-1}), alginate at 3% (w v^{-1}) and PEDOT:PSS at 3% (w v^{-1}) before and after adding PEG.

The viscosity of 3% (w v^{-1}) sodium alginate solution was roughly 8.5 Pa s, while spinning solutions of 3% (w v^{-1}) chitosan resulted in a solution with a viscosity of

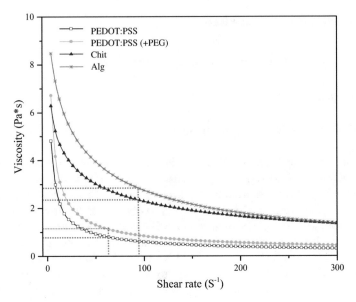

Fig. 5.3 Viscosities of spinning solutions of PEDOT:PSS with and without PEG, chitosan and sodium alginate solutions

6.4 Pa s. By increasing the shear rate, the viscosities of the two solutions became closer (The reasons for this have been discussed previously in Sect. 3.3.2). It can also be seen that by increasing the shear rate, the viscosity of PEDOT:PSS solutions generally decreased. This indicates that both solutions exhibited shear thinning behaviour. The viscosities of both gels were higher than those of PEDOT:PSS dispersions at low shear rates. It was found that the addition of 10% (w v^{-1}) PEG only increased the viscosity slightly as a result of raising the concentration. As mentioned, for the core fibre spinning, PEDOT:PSS (with or without PEG) was injected at a rate of 15 mL h^{-1} while V_i for both sheath spinning solutions was 28 mL h^{-1}. The viscosity of the 3% (w v^{-1}) PEDOT:PSS dispersion was 4.8 Pa s, while the viscosity of PEDOT:PSS (+PEG) was approximately 6.7 Pa s at shear rates close to zero. This might be considered as a benefit for coaxial wet-spinning as the viscosities of the core material became comparable to that of the sheath solutions. It has been established previously that PEG dispersions behave as Newtonian fluids below certain values of shear rates (~1000 s^{-1}) [54]. However, when the applied shear rate or shear stress exceeds a certain value, gels show shear-thickening behaviour. Shear-thickening is a time-independent non-Newtonian behaviour which is defined as the increase of viscosity with an increase in shear rate [55], often termed as a dilatant fluid. Shear thickening fluids (STFs) usually display a shear thinning behaviour at low shear stresses or shear rate. This nonlinear rheological behaviour of dispersions arises from a micro-structural rearrangement of the particles within the system. Since the applied shear rate was not in the shear-thickening regime, the dispersion has low viscosity and flows easily; thus, the presence of PEG did not cause significant changes in the viscosity of PEDOT:PSS solution in our experiment. In addition, the viscosity of all the solutions was comparable at higher shear rates.

Considering the outlet sectional area and V_i for each component (15 mL h^{-1} for PEDOT:PSS dispersion and 28 mL h^{-1} for chitosan solution), the output flow rates were calculated. The flow rates were calculated to be about ~94 s^{-1} for chitosan or alginate and ~62 s^{-1} for PEDOT:PSS (+PEG) solutions which resulted in a viscosity of ~2.4, ~2.9, ~0.95 and ~1.2 Pa s for chitosan, alginate, PEDOT:PSS and PEDOT:PSS (+PEG), respectively during the spinning process (shown in Fig. 5.3).

5.3.1.2 Impedance and Electrical Conductivity of Hydrogels

Figure 5.4a shows an example of a series of Bode plots for chitosan 3% (w v^{-1}) aqueous hydrogel in the wet gel state. The impedance magnitude (|Z|) decreases with increasing frequency and becomes independent of frequency above 1 kilohertz (kHz). This can be attributed to lower ionic diffusivity at low frequencies, due to the random path motion of ions. At higher frequencies, this effect is eliminated due to the more rapid switching of current flow direction. This infers that the system is dominated by ionic, rather than electronic, charge carriers. Figure 5.4b shows a linear relationship between the frequency-independent impedance (Z_I) and sample length; the slope of this plot was used to calculate conductivity according to Eq. 5.1.

This information was used to try and achieve optimal performance for the final battery device. The ionic conductivity of a polymer electrolyte is related to the ionic mobility and motion of polymer chains which determines their ability to continuously create a free volume into which the ions can migrate [50]. These free volumes will facilitate the diffusion of charge through the polymeric matrix, thus enhancing the ionic conductivity [56]. Enhancement of ionic conductivity will increase the flow of electrical charge between the cathode and anode and consequently improve battery performance. Electrical impedance analysis of chitosan 3% (w v^{-1}) is shown in Fig. 5.4a, b.

Table 5.1 shows the calculated electrical conductivity of hydrogels with varying salt content. As expected, the conductivity increased for increasing salt content across all chitosan and alginate hydrogels. For both chitosan and alginate hydrogels, it can be seen that 1% w v^{-1} LiCl exhibited the highest conductivity. However, NaCl at 1% (w v^{-1}) was chosen as the desired salt to be added to both hydrogels since it showed less variability (due to the smaller errors reported in Table 5.1 based on three-time measurements) as well as comparable conductivities to data from LiCl. In addition, this salt is inexpensive and abundant.

Viscosity changes versus different shear rates (0–300 s^{-1}) were determined after the addition of 1% (w v^{-1}) sodium chloride. The changes in viscosity of both hydrogels upon addition of an inert and inorganic salt are compared in Fig. 5.5 obtained from alginate and chitosan solutions with and without NaCl 1% (w v^{-1}).

It is clear from Fig. 5.5 that the viscosity of alginate and chitosan solutions both decreased by increasing the shear rate. This is an indication of shear thinning behaviour. Addition of a monovalent salt such as sodium chloride at 1% (w v^{-1})

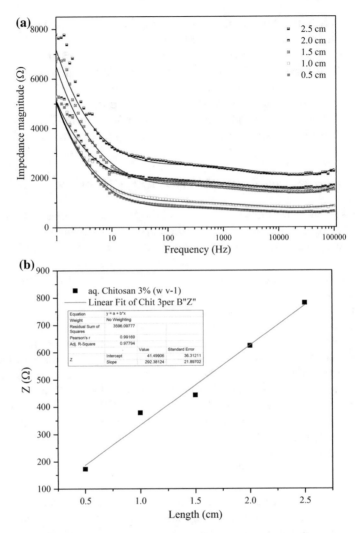

Fig. 5.4 Electrical impedance analysis of chitosan hydrogel 3% (w v^{-1}); **a** Bode plot and **b** Impedance as a function of sample length in wet-state fibres

caused a slight decrease in the viscosity of both hydrogels. It is well known that viscosity can be highly sensitive to ion type and concentration due to the way they are able to change the range and strength of the electrostatic interactions between the molecules within a sample. This phenomenon is known as electrostatic screening or screening effect [57]. Alginate and chitosan both reveal polyelectrolytic natures as they become ionised when dissolved in aqueous media. Presence of an inorganic salt invokes changes to their conformational properties [58]. When electrolytes are absent, or present only at low concentrations (dilute solutions), the polymer chains

Table 5.1 Ionic conductivity results of hydrogels using two kinds of salts

Sample name	Conductivity (m Scm^{-1})
Chit wet gel	0.38 ± **0.041**
Chit - NaCl 0.5% (w v^{-1})	11.4 ± **0.188**
Chit - NaCl 1% (w v^{-1})	15.1 ± **0.073**
Chit – LiCl 0.5% (w v^{-1})	18.9 ± **2.046**
Chit - LiCl 1% (w v^{-1})	17.5 ± **0.227**
Chit - 1M NaOH	1.1 ± **0.011**
Alg wet gel	5.8 ± **0.098**
Alg – NaCl 0.5% (w v^{-1})	14.5 ± **0.091**
Alg – NaCl 1% (w v^{-1})	22.7 ± **0.233**
Alg – LiCl 0.5% (w v^{-1})	4.8 ± **0.103**
Alg – LiCl 1% (w v^{-1})	25.4 ± **0.659**
Alg - CaCl$_2$ 2%(w v^{-1})	8.9 ± **0.219**

are more expanded due to intrachain electrostatic repulsion. However, when an inert salt is added (non-dilute solutions) a screening of the charge takes place, the electrostatic interactions decrease and the conformation of the chain becomes more compact [59]. This leads to a decrease in the viscosity of the solution when compared to salt-free conditions [60]. According to Fig. 5.5a, b, the viscosity is strongly controlled by electrostatic molecular interactions at low shear rates. However, at higher shear rates where the size and shape of the molecules become the defining feature, these interactions break down. This will be confirmed by observing the variation in viscosity as the shear rate is changed.

5.3.1.3 Microscopic Investigation of As-prepared Fibres

5.3.1.3.1. Morphological Observation in Wet-State

Figure 5.6a–c displays wet Chit-PEDOT (+PEG) coaxial fibres in water immediately after the spinning process [61] within a 1-min time period (intervals of 30 s). These images were taken from fibres which had not undergone a post-treatment procedure with propan-2-ol or isopropanol (ISP) after preparation to highlight the impact of the post-treatment on the formation of coaxial fibres. As seen from the micrographs, the conductive core is not well bound to the outer sheath without the post-treatment, and this can result in the formation of a hollow hydrogel fibre within a short time after preparation.

The same fibres were exposed to a post-treatment process straight away and imaged using the optical microscope once again. The stereomicroscopic images of wet Chit-PEDOT (+PEG) and Alg-PEDOT (+PEG) coaxial fibres are shown in Fig. 5.7a–d. The coaxial fibres showed uniform linear structures with the PEDOT:PSS

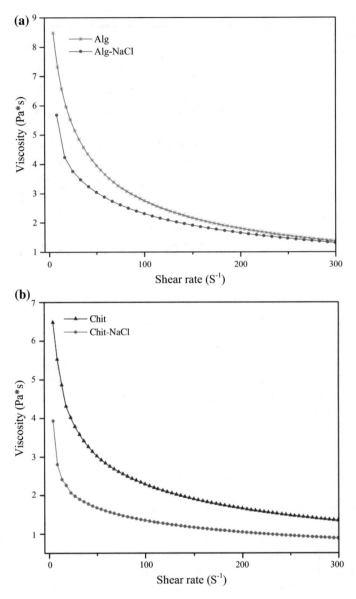

Fig. 5.5 Viscosity changes of **a** Alg and **b** chitosan spinning solutions with and without addition of NaCl 1% (w v^{-1})

(+PEG) core loaded into the fibre sheath. Moreover, the fibres appeared to remain quite stable in the media with a uniform and consistent core retained in place. The diameters of the fibres were measured to be about 580 µm for Chit-PEDOT (+PEG)

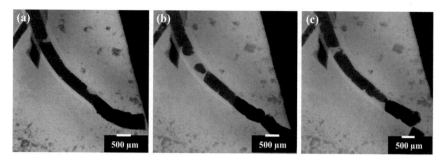

Fig. 5.6 Stereomicroscope images of (**a–c**) surface of wet Chit-PEDOT (+PEG) fibres without post-treatment

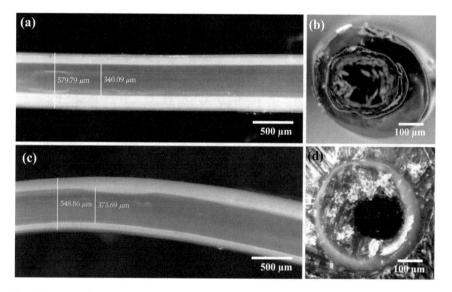

Fig. 5.7 Stereomicroscope images of surface and **b** cross-sections of (**a**), **b** Chit-PEDOT (+PEG) and **c** and **d** Alg-PEDOT (+PEG), respectively

and 550 μm for Alg-PEDOT (+PEG) fibres, in which the core was about 340 and 380 μm, respectively.

Also, the cross-sections of coaxial fibres of Chit-PEDOT were observed under LV-SEM before and after addition of PEG to the core component. Figure 5.8a–c reveals coaxial fibres in which chitosan was used as the surrounding sheath component. These images give valuable information about the internal structures of the fibres.

Cross-section of both fibre structures exhibited sealed rounded forms as clearly depicted in the LV-SEM images. The core component in Chit-PEDOT fibres in Fig. 5.8a showed a fractured, inconsistent surface. Addition of PEG into the core at 10% (w v^{-1}) resulted in creation of a crack-free and consistent central core compared to the pure PEDOT:PSS as shown in Fig. 5.8b. A porous, spongy texture is

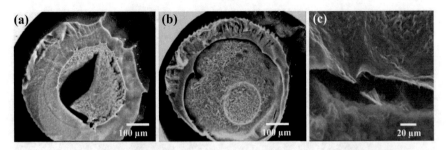

Fig. 5.8 LV-SEM images of hydrated cross sections of as-prepared **a** Chit-PEDOT, **b** Chit-PEDOT (+PEG) and **c** higher magnification of Chit-PEDOT (+PEG)

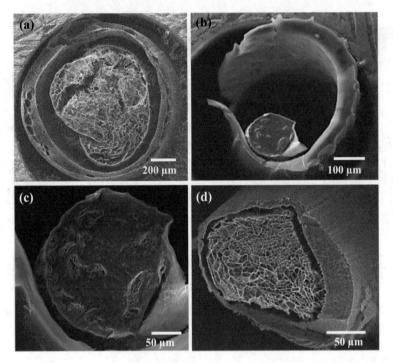

Fig. 5.9 LV-SEM images of hydrated cross sections of as-prepared **a** Alg-PEDOT, **b** Alg-PEDOT (+PEG), **c** higher magnification of Alg-PEDOT (+PEG) and **d** re-hydrated Alg-PEDOT (+PEG) fibres

noticeable in both chitosan and PEDOT sections which look as if they have reduced in size after insertion of PEG (Fig. 5.9a–b). Internal cross-section of Alg-PEDOT coaxial fibres imaged under LV-SEM clearly showed the cylinder-shaped form of the fibre as indicated in Fig. 5.9a–d. Similar to the method used for improving the consistency of the core component in regard to Chit-PEDOT fibres, PEG was also added into the PEDOT:PSS component in Alg-PEDOT fibres.

Both Alg and PEDOT constituents displayed porous permeable structures with holes visually larger than that of Chit-PEDOT (+PEG) fibres. Core and sheath components also appeared to be separated by a gap of approx. 30–50 μm. However, in contrast to the behaviour observed in Sect. 4.3.2.1 when GO was exposed to alginate, PEDOT:PSS does not show miscibility with alginate. This performance may arise from its higher viscosities (~6.7 Pa s) compared to GO solutions (~1.5 Pa s) coming from the chain entanglements present in the chemical structure due to its polymeric nature. In order to improve core and sheath attachment by building a bridge between them, PEG was added to alginate spinning solutions as well as being a common element existing in the core. However, the results revealed that addition of PEG into the core created a larger gap between the two constituents as both sections became significantly denser as shown in Fig. 5.9b. Shrinkage of the alginate sheath might be due to the creation of covalent bonding with the alginate. Insertion of PEG into alginate has been shown to regulate the swelling properties of alginate hydrogels [62]. Furthermore, addition of PEG into PEDOT:PSS, with a polymeric chemical structure, decreased the porosity of the core structure. Thus, PEG was only added to PEDOT:PSS core material as previously done for the chitosan sheath fibres. These fibres also displayed the ability to restore their initial cylindrical cross-sectional shape through several dehydration and rehydration processes as demonstrated in Fig. 5.9d.

5.3.1.3.2. Morphological Observation in Dry-State

Obtaining the right morphology is one of the most crucial steps in designing a triaxial structure. The microstructure of dehydrated Chit-PEDOT (+PEG), Alg-PEDOT (+PEG) and PPy-Chit-PEDOT (+PEG) fibres were imaged under SEM and the results are shown in Fig. 5.10a–i. As is well known, wet-spinning yields fibres of generally round or bean-shaped cross-section [63]. Although hydrated coaxial Chit-PEDOT (+PEG) and Alg-PEDOT (+PEG) fibres clearly indicated more or less disk-shaped circular cross-sections, they fully collapsed into non-regular structures after dehydration as shown in Fig. 5.10a, b. Nevertheless, a clear boundary could be found between the two components. It is also worth noting that freeze drying of fibres does not seem to be a successful method in keeping the circular cross-section. Looking at the surface patterns of coaxial fibres in Fig. 5.10e, f showed that while Chit-PEDOT (+PEG) fibres displayed a non-uniform, irregular and porous surface pattern, Alg-PEDOT (+PEG) revealed a relatively smooth surface structure with longitudinal indentations running parallel to the fibre axis. The cross-sectional shape of PPy-Chit-PEDOT (+PEG) triaxial fibres displayed an irregular structure with three clear layers shown in Fig. 5.10g–i. SEM micrographs of PPy-Chit-PEDOT (+PEG) fibres in Fig. 5.10i clearly displayed special patterns arranged in rows covering the surface of the chitosan as has also been reported for polypyrrole microstructure elsewhere [64].

To investigate the internal compositions of the cross-sections and determine if each of the layers and sections maintained their original positions after dehydration, chemical analysis was carried out through EDS mapping. This analysis could also

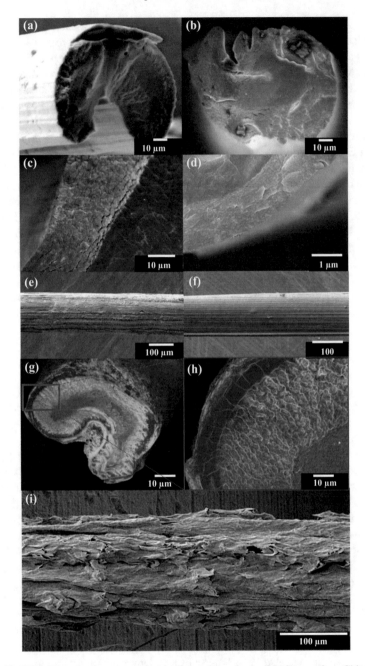

Fig. 5.10 SEM images of cross sections of as-prepared coaxial fibres of **a** Chit-PEDOT (+PEG), **b** Alg-PEDOT(+PEG) and higher magnifications of **c** Chit-PEDOT (+PEG) and **d** Alg-PEDOT(+PEG) fibre, surface pattern of **e** Chit-PEDOT (+PEG) and **f** Alg-PEDOT (+PEG) fibre, **g** cross section, **h** higher magnification of cross-section and **i** surface of PPy-Chit-PEDOT (+PEG) triaxial fibres

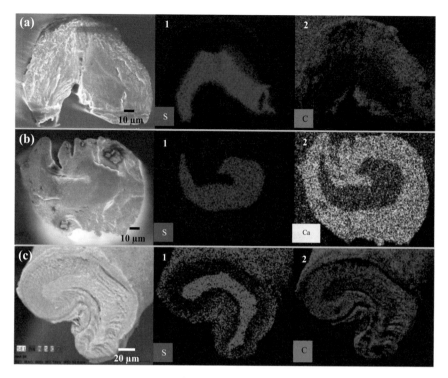

Fig. 5.11 EDS maps from cross-sections of as-prepared **a** Chit-PEDOT (+PEG), **b** Alg-PEDOT (+PEG) and **c** PPy-Chit-PEDOT (+PEG) fibres; where **a**, **b** and **c** (1) (dark blue), **a** and **c** (2) (red) and **b** (2) (light blue) shows the elemental maps of sulphur (S), carbon (C) and calcium (Ca), respectively

possibly help specify how the PEDOT (+PEG) core component is enclosed within the chitosan sheath. EDS, energy dispersive spectroscopy, relies on the interaction of the electron beam in the SEM with the specimen atoms. This interaction generates element specific X-rays which allows us to map the elemental distribution or relative abundance across a scanned image. Secondary electron images of coaxial Chit-PEDOT (+PEG), Alg-PEDOT (+PEG) and triaxial PPy-Chit-PEDOT (+PEG) fibres and their corresponding X-ray maps are depicted in Fig. 5.11a–c, respectively. Poly (3,4-ethylenedioxythiophene) sulfonate Polystyrene or PEDOT:PSS is a polymer mixture of two ionomers [65]. One component in this mixture is sodium polystyrene sulfonate. Part of the sulfonyl groups are deprotonated and carry a negative charge. The other component poly (3,4-ethylenedioxythiophene), or PEDOT, is a conjugated polymer and carries positive charges. Thus, it was determined that sulphur would be a suitable element to map as an indicator of the presence of PEDOT:PSS since it is not present in any of the hydrogels or PPy.

The EDS maps obtained clearly confirmed the results previously obtained from SEM investigations. The bean structured cross-sections of coaxial and triaxial fibres

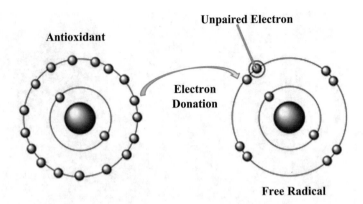

Fig. 5.12 Mechanism of how antioxidants reduce free radicals by giving the free radical an electron which inhibits other oxidation reactions

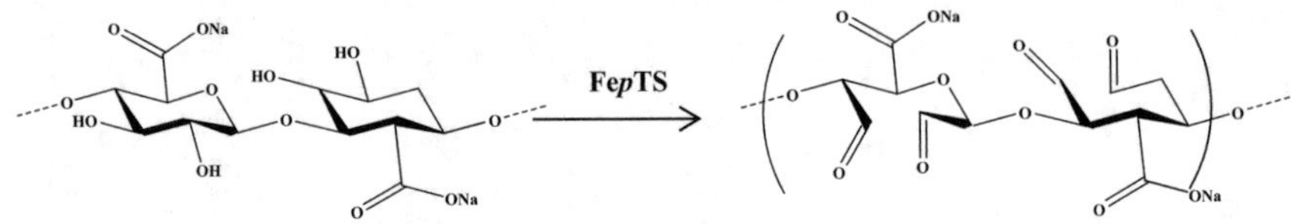

Fig. 5.13 An oxidation reaction of alginate upon addition of the oxidant

were shown to enclose the sulphur-rich conductive core of PEDOT:PSS. A thin layer of PPy was also observed to be deposited as the third layer on the surface of Chit-PEDOT (+PEG) fibres (Fig. 5.11c)

As noted, the polymerisation of pyrrole could not be initiated on the surface of alginate it was previously for chitosan. To investigate the reasons for this we can possibly look back at the chemical structures of both alginate and chitosan. As discussed earlier, chitosan is a cationic polysaccharide containing amino groups which could be protonated in aqueous media [27]. Thus, chitosan would be able to display antioxidative activity due to the existence of these lone so-called "unpaired electrons" in its structure. The damaging process that is prevented by antioxidants is the prevention of free radicals. Consequently, it could prevent the oxidation of other chemicals from occurring [66]. A schematic shows antioxidant activity of chitosan is presented in Fig. 5.12.

Sodium alginate, on the other hand, in the absence of antioxidant protection is predicted to undergo an oxidation process when exposed to the oxidant which results in backbone cleavage [67]. In the oxidation reaction, hydroxyl groups on carbons 2 and 3 of repetitive units were oxidized by Fe*p*TS leading to the formation of bis-aldehyde functionalities in each oxidized monomeric unit by cleaving the carbon–carbon bond as shown in Fig. 5.13 [66].

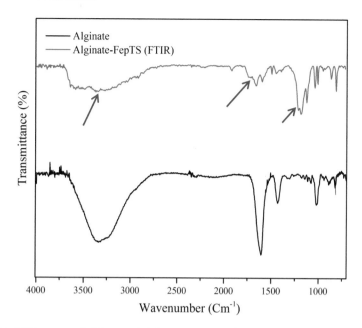

Fig. 5.14 FTIR spectra of alginate and alginate-Fe*p*TS

FTIR spectroscopy was used to evaluate the structural changes in the alginate backbone after the oxidation procedure. Consequently, Fig. 5.14 shows the resulting FTIR spectra of pure alginate (Alginate) and oxidized alginate (Alginate-Fe*p*TS).

As expected, pure alginate displayed the characteristic absorption bands of alginic acid, including the peaks at around 1603 and 1413 cm^{-1}, corresponding to asymmetric and symmetric stretching vibration of carboxylate salt groups of alginate, a peak at around 1033 cm^{-1} assigned to C–O–C stretching vibration and a peak at 3255 cm^{-1} attributed to –OH stretching vibration [68]. Interestingly, oxidized alginate showed a completely different absorption pattern from pure alginate, supporting the theory of structural substitution of functional groups due to chemical reaction between sodium alginate and the oxidizing agent (Fe*p*TS). Subsequently, a dramatic decrease in intensity of the –OH stretching vibration (at 3221 cm^{-1}) of alginate was due to oxidation at the hydroxyl groups. Identifying the newly formed aldehyde groups (as a consequence of oxidation) proved to be difficult with only a weak absorption band at 1736 cm^{-1} assigned to symmetric vibration of aldehyde (which was due to hemiacetal formation of free aldehyde groups confirmed by observation of the characteristic absorption band of hemiadol structure at 866 cm^{-1}) [66, 69–71]. Additionally, the appearance of the band at around 1222 cm^{-1} corresponds to S=O stretching vibration of the oxidizing agent (Fe*p*TS) [72, 73]. It is also noteworthy that the significant reduction in absorption bands of carboxylate groups (1603 and 1413 cm^{-1}) was due to the shielding effect of iron ions present in the oxidizing agent (Fe*p*TS) [74, 75].

Table 5.2 Mechanical properties of coaxial and triaxial fibres

Sample name	Average Young's modulus (GPa)	Ultimate stress (MPa)	Ultimate strain (%)
PEDOT:PSS fibre [12]	3.3 ± 0.3	125 ± 7	15.8 ± 1.2
Chit-PEDOT fibre	3.9 ± 0.1	94.06 ± 3.4	8 ± 1.98
Chit-PEDOT (+PEG) fibre	5.16 ± 0.6	160.25 ± 16.8	7.52 ± 0.88
Alg-PEDOT(+PEG) fibre	4.05 ± 0.08	94.81 ± 5.3	16.73 ± 1.81
PPy- Chit-PEDOT (+PEG) fibre	2.2 ± 0.25	64.81 ± 3.3	3.5 ± 0.25

5.3.1.4 Mechanical Properties

Tensile tests were carried out on a Dynamic Mechanical Tester (EZ-L Tester from Shimadzu, Japan), at 2 mm min^{-1} and a gauge length of 11.5 mm. Average values of tensile strength and maximum strain were determined for four different coaxial fibre types with each test being repeated four times. The results are summarized in Table 5.2.

Results obtained from stress–strain curves for as-spun coaxial and triaxial fibres showed a significant increase in robustness for them compared to the results previously reported for PEDOT:PSS fibres by others [12]. Young's moduli of fibres were calculated to be ~3.9, ~5.1, ~4 and ~2.2 GPa for Chit-PEDOT, Chit-PEDOT (+PEG), Alg-PEDOT (+PEG) and PPy- Chit-PEDOT (+PEG) fibres, respectively. The results indicate that the presence of a hydrogel sheath plays a major role in increasing the Young's modulus. On the other hand, deposition of the third layer has decreased the Young's modulus. Analysis of this data also indicated an ultimate stress of ~125 MPa with ~15.8% strain for solid fibres [76] compared with ~94 MPa with 8% strain, ~160 MPa with 7.5% strain and 94 MPa stress with ~16.7% for Chit-PEDOT, Chit-PEDOT (+PEG) and Alg-PEDOT(+PEG) fibres, respectively. In other words, there is a loss of ~25% of tensile strain in hybrid Chit-PEDOT fibres compared to that of solid PEDOT:PSS fibres which might be compensated for by the addition of PEG into the core of coaxial Chi-PEDOT fibres. This outcome can be explained by the creation of a strong homogenous interfacial connection between the core and sheath sections in the presence of PEG which causes efficient stress transfer when strain is applied on a tensile specimen. Moreover, Chit-PEDOT fibres were shown to withstand less stress before breakage in the elastic regime, while Alg-PEDOT (+PEG) revealed higher strain percentages before fracture. Triaxial fibres of PPy- Chit-PEDOT (+PEG) fibre were shown to have ultimate tensile stress of ~64 MPa with ~3.5% elongation at breakage. The creation of multilayered fibres with different sections allowing different properties, as well as their dissimilar responses to the applied stress, might be the reason behind this reduction. These data were calculated assuming that the cross-sectional area was circular with a diameter equal to the widest diameter of the irregular fibre.

5.3.1.5 Cyclic Voltammetry (CV)

CV curves of as-prepared coaxial fibres were collected to evaluate the EC properties of fibres. Curves for solid PEDOT:PSS(PEG), as well as coaxial Chit-PEDOT (+PEG) and Alg-PEDOT (+PEG) fibres, are shown in Fig. 5.15a–b in two electrolytes, PBS and 0.1 M TBABF$_4$ (CV of cotton-steel fibre has been added as the reference). To make the connection to the PEDOT (+PEG) core of the fibres, a cotton-steel wire with an average diameter of 20 μm was inserted into the fibre core while spinning. The applied υ was 10 mV s^{-1} and 100 cycles were performed. Reasonable electroactivity is observed in CV results from coaxial fibres which also indicated a stable behaviour after 100 cycles in both media. The EC performance of individual PEDOT:PSS fibres has been previously studied in a three-electrode cell using different electrolytes [77].

The capacitive behaviour of PEDOT:PSS (+PEG) was represented by CVs showing near-rectangular shaped behaviour in aqueous and organic media in a three-electrode configuration. The near-rectangular shape of the CVs obtained from PEDOT:PSS (+PEG) suggests that the overall internal resistance is low, owing to the high fibre electroactivity. It is worth mentioning that the cotton-steel wire showed negligible electroactivity as observed in the Fig. 5.15 in both media.

PEDOT:PSS(+PEG) fibres exhibited more pronounced oxidations (−0.4 and +0.4 V vs. Ag/Ag$^+$) and reduction (−0.2 and −0.5 V vs. Ag/Ag$^+$) peaks. However, PEDOT:PSS (+PEG) fibres exhibited a well-defined EC behaviour corresponding to oxidation (−0.3 V) and reduction (−0.5 V) of the PEDOT backbone in the organic electrolyte. The oxidation and reduction peaks shifted to (+0.2 V vs. Ag/Ag$^+$) and (+0.6 V vs. Ag/Ag$^+$) for Chit-PEDOT (+PEG) and (−0.3 V vs. Ag/Ag$^+$) and (+0.5 V vs. Ag/Ag$^+$) for Alg-PEDOT (+PEG) fibres, respectively. These redox values for coaxial fibres are at potentials higher than the oxidation state of the PEDOT:PSS fibre core. This provides the possibility of switching between oxidised and reduced states for the coaxial fibres whilst the PEDOT:PSS layer remains in its oxidised (conducting) state. In addition, the CV curves showed obviously larger CV areas, and correspondingly, higher specific capacitances in 0.1 M TBABF$_4$ than those of the CVs in PBS solution. The coaxial fibres displayed a kind of tilted CV shape. Specific capacitances of 19 ± 0.65, 11 ± 1.2 and 9.2 ± 0.8 F g^{-1} (at a υ of 10 mV s^{-1}) were calculated for PEDOT:PSS (+PEG), Chit-PEDOT:PSS (+PEG) and Alg-PEDOT:PSS (+PEG) fibres, respectively, in PBS solution (based on the calculations in Sect. 2.2.1.11). Much higher capacitances of 29.7 ± 1.5, 21 ± 0.2 and 13.05 ± 0.6 F g^{-1} were measured for PEDOT:PSS (+PEG), Chit-PEDOT:PSS (+PEG) and Alg-PEDOT:PSS (+PEG) fibres, respectively in 0.1 M TBABF$_4$.

It was expected that we would see higher electroactivities from Alg-PEDOT (+PEG) fibres compared to that of Chit-PEDOT (+PEG) because of their higher ionic conductivities in wet-state (~5.8 for alginate vs. ~0.38 for chitosan). Alg-PEDOT (+PEG) fibres, however, demonstrated less electroactivity compared to Chit-PEDOT (+PEG) fibres as shown by the lower current measured in CV curves. This outcome

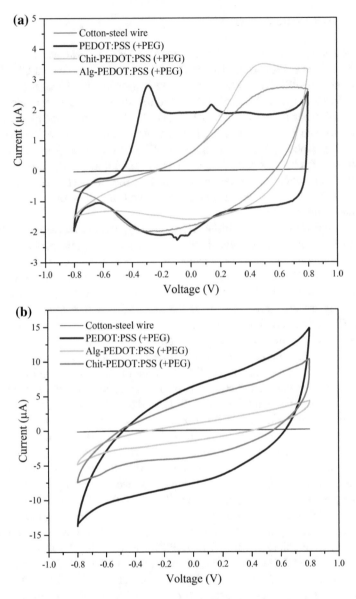

Fig. 5.15 Cyclic voltammograms of PEDOT:PSS (+PEG), Chit-PEDOT (+PEG) and Alg-PEDOT (+PEG) fibres in **a** PBS solution and **b** 0.1 M TBABF$_4$; potential was scanned between -0.8 V and $+0.8$ at 10 mV s^{-1}

might be explained by the lack of any bonding between alginate and PEDOT:PSS in the core (and is supported by the LV-SEM images in Sect. 5.3.1.3.1 which show a gap of approx. 30–50 μm).

5.3.1.6 In Vitro Bioactivity Experiments

Tests for cytotoxicity and cell adhesion are essential in assessing tissue—biomaterial compatibility integration. It is well known that the quality of cell attachment to the substrate material will determine the capacity of cells to proliferate and to differentiate into biomaterials [78, 79]. To this end, we tested cell toxicity and biocompatibility of coaxial fibres using primary cells without the use of any extra cellular matrices, such as laminin or collagen, to assess the true affinity of these cells to our synthetic substrates.

The proportion of live cells was seen to be very high on both coaxial fibre types. This was an indication that the fibres did not release any toxic chemicals that caused cell death, indicating that the materials are biocompatible. The tests revealed that on coaxial Chit-PEDOT (+PEG) fibres the primary cells attached, differentiated and proliferated. Also, the cell morphology after 5 days on the fibres was typical of primary cells. Nevertheless, it was also observed that cells did not show any tendencies toward attachment to coaxial fibres coated with alginate and attached and differentiated much better on coaxial Chit-PEDOT (+PEG) fibres than on the Alg-PEDOT (+PEG) ones as shown in Fig. 5.16. Furthermore, it was observed, that in the case of the primary cells, the density of cells growing on the fibre surfaces (shown by the calcein-stained-green colour) was slightly lower than that observed on the underlying substrate (image not shown). Overall, the myoblast cells were observed to adhere and the density of cells on Chit-PEDOT (+PEG) fibres increased over time which, together with the high cell viability, indicates that the fibres were capable of supporting myoblast cell adhesion and proliferation.

5.4 Conclusion

Biopolymeric continuous core-sheath fibres, with an inner core of PEDOT:PSS and either chitosan or alginate as the sheath, were fabricated for the first time without using a template *via* a simple wet-spinning process. Using a CVD method, a second thin conductive layer of PPy was grown on the coaxial fibre surface leading to the production of triaxial fibres to be utilized as potential microscale biobatteries. The morphological, mechanical, electrochemical and biological properties of these fibres are discussed. Cross-sectional images of coaxial fibres taken by LV-SEM also confirmed the role of PEG in the creation of a crack-free and consistent core texture compared to pure PEDOT:PSS. Mechanical property results obtained from stress–strain curves for as-spun coaxial fibres showed a significant increase in robustness for them compared to results previously reported for PEDOT:PSS fibres by others. Sodium chloride at 1% (w v^{-1}) was chosen as the desired salt solution to be added to the electrolyte layer, giving the highest ionic conductivity in chitosan and alginate with lower standard deviations compared to lithium chloride. Furthermore, reasonable electroactivity was observed in CV results from coaxial fibres which also indicated stable behaviour in both media. In vitro bioactivity experiments suggested

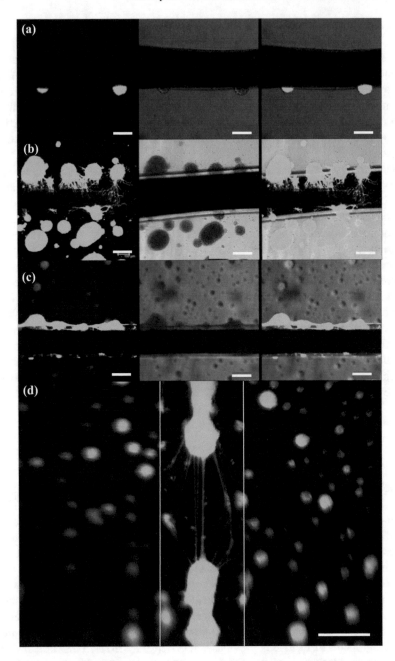

Fig. 5.16 Primary cell attachment and proliferation on fibres after 5 days of differentiation. **a** Alg-PEDOT (+PEG) and Chit-PEDOT (+PEG) fibres from **b** bottom **c** side and **d** top view, respectively. Scale bars represent 100 μm

that the fibres did not release any toxic component that caused cell death; however, primary cell lines appeared to attach and differentiate much better onto coaxial Chit-PEDOT (+PEG) fibres compared to those coated with an alginate sheath. All prepared fibres showed good electrochemical and mechanical properties, as well as cytocompatiblity, which make them useful for potential applications as biosensors, electrodes, tissue scaffolds or biobatteries.

References

1. Stoppa M, Chiolerio A (2014) Wearable electronics and smart textiles: a critical review. Sensors 14:11957–11992. https://doi.org/10.3390/s140711957
2. Seyedin S, Razal JM, Innis PC, Jeiranikhameneh A, Beirne S, Wallace GG (2015) Knitted strain sensor textiles of highly conductive all-polymeric fibers. ACS Appl Mater Interfaces 7:21150–21158. https://doi.org/10.1021/acsami.5b04892
3. Davy J (2012) Advances in military textiles and personal equipment. Woodhead Publishing Limited
4. Jia X, Yang Y, Wang C, Zhao C, Vijayaraghavan R, Macfarlane DR, Forsyth M, Wallace GG (2014) Biocompatible ionic liquid–biopolymer electrolyte-enabled thin and compact magnesium–air batteries. Appl Mater Interfaces 6:21110–21117
5. Tao J, Liu N, Ma W, Ding L, Li L, Su J, Gao Y (2013) Solid-state high performance flexible supercapacitors based on polypyrrole-MnO$_2$-carbon fiber hybrid structure. Sci Rep 3:1–7. https://doi.org/10.1038/srep02286
6. Nanofiber GP, Wu Q, Xu Y, Yao Z, Liu A, Shi G (2010) Supercapacitors based on flexible-graphene/polyaniline nanofiber composite films. ACS Nano 4:1963–1970
7. Shim BS, Chen W, Doty C, Xu C, Kotov NA (2008) Smart electronic yarns and wearable fabrics for human biomonitoring made by carbon nanotube coating with polyelectrolytes. Nano Lett 8:4151–4157
8. Liu J, Li N, Goodman MD, Zhang HG, Epstein ES, Huang B, Pan Z, Kim J, Choi JH, Huang X, Liu J, Hsia KJ, Dillon SJ, Braun PV (2015) Mechanically and chemically robust sandwich-structured C @ Si @ C nanotube array li-ion battery anodes. ACS Nano 9:1985–1994
9. Gaikwad AM, Steingart DA, Ng TN, Schwartz DE, Whiting GL (2013) A flexible high potential printed battery for powering printed electronics. Appl Phys Lett 102:233302-1-5. https://doi.org/10.1063/1.4810974
10. Wang J, Too CO, Zhou D, Wallace GG (2005) Novel electrode substrates for rechargeable lithium/polypyrrole batteries. J Power Sources 140:162–167. https://doi.org/10.1016/j.jpowsour.2004.08.040
11. Wei L, Zhang K, Tao Z, Chen J (2015) Sn–Al core—shell nanocomposite as thin film anode for lithium-ion batteries. J Alloy Compod 644:742–749
12. Jalili R, Razal JM, Innis PC, Wallace GG (2011) One-step wet-spinning process of poly (3, 4-ethylenedioxy- thiophene): poly (styrenesulfonate) fibers and the origin of higher electrical conductivity. Adv Func Mater 21:3363–3370. https://doi.org/10.1002/adfm.201100785
13. Liu Y, Li X, Lu JC (2013) Electrically conductive poly (3, 4-ethylenedioxythiophene)–polystyrene sulfonic acid/polyacrylonitrile composite fibers prepared by wet spinning. J Appl Polym Sci 130:370–374. https://doi.org/10.1002/app.39174
14. Zampetti E, Macagnano A, Pantalei S, Bearzotti A (2012) PEDOT:PSS coated titania nanofibers for NO2 detection: Study of humidity effects. Sensors Actuators B Chem. https://doi.org/10.1016/j.snb.2012.09.107
15. Esrafilzadeh D, Razal J, Moulton S, Stewart E, Wallace G (2013) Multifunctional conducting fibres with electrically controlled release of ciprofloxacin. J Control Release 169:313–320

16. Okuzaki H, Harashina Y, Yan HH (2009) Highly conductive PEDOT/PSS microfibers fabricated by wet-spinning and dip-treatment in ethylene glycol. Eur Polym J 45:256–261. https://doi.or g/10.1016/j.eurpolymj.2008.10.027
17. Gao Y, Li J, Yang X, Xiang Q, Wang K (2014) Electrochemiluminescence biosensor based on PEDOT-PSS- graphene functionalized ITO electrode. Electroanalysis 26:382–388. https://do i.org/10.1002/elan.201300470
18. Moczko E, Istamboulie G, Calas-Blanchard C, Rouillon R, Noguer T (2012) Biosensor employing screen-printed PEDOT:PSS for sensitive detection of phenolic compounds in water. J Polym SciA 50:2286–2292. https://doi.org/10.1002/pola.26009
19. Weis M, Otsuka T, Taguchi D, Manaka T, Iwamoto M (2015) Charge injection and accumulation in organic light-emitting diode with PEDOT:PSS anode. J Appl Phys 117. https://doi.org/10.1 063/1.4918556
20. Hooper KEA, Smith B, Greenwood P, Baker J, Watson TM (2015) Spray PEDOT:PSS coated perovskite with a transparent conducting electrode for low cost scalable photovoltaic devices. Mater Res Innov 19:482–487. https://doi.org/10.1080/14328917.2015.1105572
21. Kim HP, Lee SJ, Jang J (2015) Improvement of conversion efficiency of inverted organic photovoltaic with PEDOT: PSS: WOx by thermal annealing. IEEE J PHOTOVOLTAICS 5:897–902
22. Okuzaki H, Ishihara M (2003) Spinning and characterization of conducting microfibers. Macromol Rapid Commun 24:261–264
23. Kumar M (1999) Chitin and chitosan fibres: a review. Bull Mater Sci 22:905–915
24. Majima T, Funakosi T, Iwasaki N, Yamane S-TT, Harada K, Nonaka S, Minami A, Nishimura S-II (2005) Alginate and chitosan polyion complex hybrid fibers for scaffolds in ligament and tendon tissue engineering. J Orthop Sci 10:302–307. https://doi.org/10.1007/s00776-005-089 1-y
25. Niekraszewicz A (2005) Chitosan medical dressings. Fibres Text East Eur 13:16–18
26. Prabaharan M, Mano JF (2005) Chitosan-based particles as controlled drug delivery systems. Drug Deliv 12:41–57. https://doi.org/10.1080/10717540590889781
27. Wan Y, Creber KAM, Peppley B, Bui VT (2003) Ionic conductivity of chitosan membranes. polymer (Guildf) 44:1057–1065
28. Wang L, Khor E, Wee A, Lim LY (2002) Chitosan-Alginate PEC membrane as a wound dressing: assessment of incisional wound healing. J Biomed Mater Res 63:610–618. https://d oi.org/10.1002/jbm.10382
29. Cui CJ, Wu GM, Yang HY, She SF, Shen J, Zhou B, Zhang ZH (2010) A new high-performance cathode material for rechargeable lithium-ion batteries: polypyrrole/vanadium oxide nanotubes. Electrochim Acta 55:8870–8875. https://doi.org/10.1016/j.electacta.2010.07.087
30. Kakuda S, Momma T, Osaka T (1995) Ambient-temperature, rechargeable, all-solid lithium/polypyrrole polymer battery. J Electrochem Soc 142:1–2
31. Huang ZB, Yin GF, Liao XM, Gu JW (2014) Conducting polypyrrole in tissue engineering applications. Front Mater Sci 8:39–45. https://doi.org/10.1007/s11706-014-0238-8
32. Jager EWH, Immerstrand C, Magnusson K, Inganas O, Lundstrom I (2000) Biomedical applications of polypyrrole mieroactuators : from single-cell clinic to microrobots. In: Annual International IEEE-EMBS Special Topic Conference on Microtechnologies in Medicine & Biology. pp 58–61
33. Svirskis D, Travas-sejdic J, Rodgers A, Garg S (2009) Polypyrrole film as a drug delivery system for the controlled release of risperidone. In: AIP Conference Proceeding. pp 36–40
34. Svirskis D, Wright BE, Travas-sejdic J, Rodgers A, Garg S (2010) Sensors and actuators B: chemical evaluation of physical properties and performance over time of an actuating polypyrrole based drug delivery system. Sens Actuators, B 151:97–102. https://doi.org/10.1016/j.snb. 2010.09.042
35. Hamilton S, Hepher MJ, Sommerville J (2005) Polypyrrole materials for detection and discrimination of volatile organic compounds. Sens Actuators, B 107:424–432. https://doi.org/1 0.1016/j.snb.2004.11.001
36. Qin H, Kulkarni A, Zhang H, Kim H, Jiang D, Kim T (2011) Polypyrrole thin film fiber optic chemical sensor for detection of VOCs. Sens Actuators, B 158:223–228. https://doi.org/10.10 16/j.snb.2011.06.009

37. Foroughi J, Spinks GM, Wallace GG, Whitten PG (2008) Production of polypyrrole fibres by wet spinning. Synth Met 158:104–107. https://doi.org/10.1016/j.synthmet.2007.12.008
38. Rowley NM, Mortimer RJ (2002) New electrochromic materials. Sci Prog 85:243–262
39. Cho JW, Jung H (1997) Electrically conducting high-strength aramid composite fibres prepared by vapour-phase polymerization of pyrrole. J Mater Sci 32:5371–5376
40. Lawal AT, Wallace GG (2014) Vapour phase polymerisation of conducting and non-conducting polymers: a review. Talanta 119:133–143. https://doi.org/10.1016/j.talanta.2013.10.023
41. Xu C, Wang P, Bi X (1995) Continuous vapor phase polymerization of pyrrole. I. electrically conductive composite fiber of polypyrrole with poly(p-phenylene terephthalamide). J Appl Polym Sci 58:2155–2159
42. Yang Y, Zhang L, Li S, Wang Z, Xu J, Yang W, Jiang Y (2013) Vapor phase polymerization deposition conducting polymer nanocomposites on porous dielectric surface as high performance electrode materials. Nano-Micro Lett 5:40–46
43. Ateh D, Navsaria H, Vadgama P (2006) Polypyrrole-based conducting polymers and interactions with biological tissues. J R Soc Interface 3:741–752. https://doi.org/10.1098/rsif.2006.0141
44. Chen Liangbi, Chen Wenfeng, Ma Chunhua, Du Dan XC (2011) Electropolymerized multi-walled carbon nanotubes-polypyrrole fiber for solid-phase microextraction and its applications in the determination of pyrethroids. Talanta 84:104–108
45. Maziz A, Khaldi A, Persson N, Jager EWH (2015) Soft linear electroactive polymer actuators based on polypyrrole. In: Proc. of SPIE. pp 1–6
46. Granero BAJ, Razal JM, Wallace GG (2008) Spinning carbon nanotube-gel fibers using poly-electrolyte complexation. Adv Func Mater 18:3759–3764. https://doi.org/10.1002/adfm.200800847
47. Tian M, Hu X, Qu L, Zhu S, Sun Y, Han G (2016) Versatile and ductile cotton fabric achieved via layer-by-layer self-assembly by consecutive adsorption of graphene doped PEDOT: PSS and chitosan. Carbon N Y 96:1166–1174. https://doi.org/10.1016/j.carbon.2015.10.080
48. Jančiauskaitė U, Višnevskij Č, Radzevičius K, Makuška R (2009) Polyampholytes from natural building blocks: synthesis and properties of chitosan-o-alginate copolymers. Chemija 20:128–135
49. Yu D, Branford-White K, Chatterton N, Zhu L, Huang L, Wang B (2011) A modified coaxial electrospinning for preparing fibers from a high concentration polymer solution. Express Polym Lett 5:732–741. https://doi.org/10.3144/expresspolymlett.2011.71
50. Warren H, Gately RD, Brien PO, Iii RG (2014) Electrical conductivity, impedance, and per-colation behavior of carbon nanofiber and carbon nanotube containing gellan gum hydrogels. J Polym Phys 52:864–871. https://doi.org/10.1002/polb.23497
51. Jayakumar R, Prabaharan M, Kumar P, Nair S, Tamura H (2011) Biomaterials based on chitin and chitosan in wound dressing applications. Biotechnol Adv 29:322–337. https://doi.org/10.1016/j.biotechadv.2011.01.005
52. Shao X, Hunter CJC (2007) Developing an alginate/chitosan hybrid fiber scaffold for annulus fibrosus cells. J Biomed Mater Res A 82:702–710. https://doi.org/10.1002/jbm.a
53. Chem JM, Jalili R, Razal JM, Wallace GG (2012) Exploiting high quality PEDOT:PSS–SWNT composite formulations for wet-spinning multifunctional fibers. J Mater Chem 22:25174–25182. https://doi.org/10.1039/c2jm35148j
54. Brikov A, Markin A, Sukhoverkhov S (2015) Rheological properties of polyethylene glycol solutions and gels. Ind Chem 1:1–5. https://doi.org/10.4172/2469-9764.1000102
55. Hadley DW (1975) Rheological nomenclature. Rheol acta 14:1098–1109. https://doi.org/10.1007/BF01515905
56. Christie AM, Lilley SJ, Staunton E, Andreev YG (2005) Increasing the conductivity of crys-talline polymer electrolytes. Nature 433:50–53. https://doi.org/10.1038/nature03190.1
57. Sadeghi M, Ghasemi N (2012) Synthesis and study on effect of various chemical conditions on the swelling property of collagen-g-poly (AA- co -IA) superabsorbent hydrogel. Indian J Sci Technol 5:1879–1884

58. Wang X, Li Y, Li J, Wang J, Wang Y, Guo Z (2005) salt effect on the complex formation between polyelectrolyte and oppositely charged surfactant in aqueous solution. J Phys Chem B 109:10807–10812

59. Dyakonova MA, Berezkin AV, Kyriakos K, Gkermpoura S, Popescu MT, Filippov SK, Petr S, Papadakis CM (2015) Salt-induced changes in triblock polyampholyte hydrogels: computer simulations and rheological, structural, and dynamic characterization. Macromol 48:8177–8189. https://doi.org/10.1021/acs.macromol.5b01746

60. Pamies R, Schmidt RR, López C, García J, Torre D (2010) The influence of mono and divalent cations on dilute and non-dilute aqueous solutions of sodium alginates. Carbohyd Polym 80:248–253. https://doi.org/10.1016/j.carbpol.2009.11.020

61. Tangsadthakun C, Kanokpanont S, Sanchavanakit N, Banaprasert T, Damrongsakkul S (2006) Properties of collagen/chitosan scaffolds for skin tissue engineering fabrication of collagen/chitosan scaffolds. J Met Mater Miner 16:37–44

62. Lee KY, Mooney DJ (2012) Alginate: properties and biomedical applications. Prog Polym Sci 37:106–126. https://doi.org/10.1016/j.progpolymsci.2011.06.003

63. Cook G (2001) Polyvinyl derivatives. In: Handbook of textile fibres man-made fibres. pp 392–535

64. Grzeszczuk M, Ozsakarya R (2014) Surface morphology and corresponding electrochemistry of polypyrrole films electrodeposited using a water miscible ionic liquid. RSC Adv 4:22214–22223. https://doi.org/10.1039/c4ra03497j

65. Irwin MD, Roberson DA, Olivas RI, Wicker RB, Macdonald E (2011) Conductive polymer-coated threads as electrical interconnects in e-textiles. Fibers Polym 12:904–910. https://doi.org/10.1007/s12221-011-0904-8

66. Sarker B, Papageorgiou DG, Silva R, Zehnder T, Gul-e-noor F, Bertmer M, Kaschta J, Detsch R, Boccaccini AR (2014) Fabrication of alginate—gelatin crosslinked hydrogel microcapsules and evaluation of the microstructure and physico-chemical properties. J Mater Chem B 2:1470–1482. https://doi.org/10.1039/c3tb21509a

67. Boontheekul T, Kong H, Mooney DJ (2005) Controlling alginate gel degradation utilizing partial oxidation and bimodal molecular weight distribution. Biomaterials 26:2455–2465. https://doi.org/10.1016/j.biomaterials.2004.06.044

68. Lawrie G, Keen I, Drew B, Chandler-temple A, Rintoul L, Fredericks P, Grøndahl L (2007) Interactions between Alginate and Chitosan Biopolymers Characterized Using FTIR and XPS. Biomacromol 8:2533–2541

69. Naderi A, Lindstr T (2015) One-shot carboxylation of microcrystalline cellulose in the presence of nitroxyl radicals and sodium periodate. RSC Adv 5:85889–85897. https://doi.org/10.1039/c5ra16183e

70. Tan H, Chu CR, Payne KA, Marra KG (2009) Injectable in situ forming biodegradable chitosan—hyaluronic acid based hydrogels for cartilage tissue engineering. Biomaterials 30:2499–2506. https://doi.org/10.1016/j.biomaterials.2008.12.080

71. Wang X, Gu Z, Qin H, Li L, Yang X, Yu X (2015) Crosslinking effect of dialdehyde starch (DAS) on decellularized porcine aortas for tissue engineering. Int J Biol Macromol 79:813–821

72. Galligani M, Rondón RA, Ovalles JF, Brunetto MR (2014) Transmission FTIR derivative spectroscopy for estimation of furosemide in raw material and tablet dosage form. Acta Pharm Sin B 4:376–383

73. Puvaneswary S, Talebian S, Balaji H, Raman M, Mehrali M, Muhammad A, Hayaty N, Abu B, Kamarul T (2015) Fabrication and in vitro biological activity of BTCP-Chitosan-Fucoidan composite for bone tissue engineering. Carbohydr Polym 134:799–807

74. Kumar A, Bahadur R (2014) Iron crosslinked alginate as novel nanosorbents for removal of arsenic ions and bacteriological contamination from water. J Mater Res Technol 3:195–202

75. Swamy BY, Yun Y (2015) In vitro release of metformin from iron (III) cross-linked alginate—carboxymethyl cellulose hydrogel beads. Int J Biol Macromol 77:114–119

76. Jalili R, Razal JM, Wallace GG (2013) Wet-spinning of PEDOT:PSS/ functionalized-SWNTs composite: a facile route toward production of strong and highly conducting multifunctional fibers. Nature 3:1–7. https://doi.org/10.1038/srep03438

77. Jalili R (2012) Wet-spinning of nanostructured fibres. University of Wollongong
78. Anselme K (2000) Osteoblast adhesion on biomaterials. Biomaterials 21:667–681
79. Mirabedini A, Foroughi J, Thompson B, Wallace GG (2015) Fabrication of coaxial wet-spun graphene—chitosan bio fibers. Adv Eng Mater 18:284–293. https://doi.org/10.1002/adem.20 1500201

Chapter 6
Conclusion and Future Work

6.1 General Conclusion

The main objective of this thesis was to develop 3D electroactive fibres by means of the coaxial wet-spinning method to be utilized in a variety of bioapplications such as implantable electrodes, drug delivery systems as well as more complicated applications such as energy-storage systems. In order to fabricate coaxial electroactive fibres, a number of electrically conducting materials, i.e. PEDOT:PSS, LC GO and PPy, as well as biocompatible ionically conducting polymers including chitosan and alginate have been employed. The use of non-conducting materials as the surrounding sheath improves cell adhesion properties since they provide mechanical and structural properties that mimic many tissues and their extracellular matrix. However, the conducting elements in the core are necessary for the creation of electrical pathways required to allow electrical stimulation of cells. As a result of this research, the production of continuous multiaxial fibres has been achieved using a combination of the various above-mentioned materials. The focus of this thesis is on developing and characterising coaxial biofibres using naturally occurring hydrogels and organic conductors with a view to their ultimate use as biocompatible electrodes.

Wet-spinning, as the most versatile and viable fibre fabrication method, was the key approach used to develop the coaxial conducting fibres. The major difference in the coaxial wet-spinning method used through this study compared to conventional wet-spinning was that in the coaxial process, two different polymer solutions are injected into a coaxial spinneret together and are co-extruded into a bath while retaining a coaxial structure. The simple processability of both alginate and chitosan together with their biocompatibility, suitable mechanical properties as well as their ability to form polyanion–polycation complexes led to their selection as the hydrogels for use in this thesis.

Since a number of parallel solution and processing parameters are required to be optimised and controlled in order to form coaxial fibres in a coagulation bath, and also because of the limited number of existing studies in the area of effective fabrication of

© Springer Nature Switzerland AG 2018
A. Mirabedini, *Developing Novel Spinning Methods to Fabricate Continuous Multifunctional Fibres for Bioapplications*, Springer Theses,
https://doi.org/10.1007/978-3-319-95378-6_6

either coaxial or triaxial fibres *via* coaxial wet-spinning, it was necessary to initially investigate the optimum conditions for wet-spinning of hydrogel-based coaxial fibres. This was carried out and the results are presented in Chap. 3.

Fabrication of coaxial Chit/Alg fibres (inner core of chitosan and alginate as the outer sheath) was made possible using coaxial wet-spinning employing a novel method of blending the chitosan core with various percentages of calcium chloride (0.5, 1 and 2% w v^{-1}). Selection of a matching solvent, as well as the appropriate injection rates for both core and sheath components, were other critical factors that were successfully addressed throughout the thesis. The results showed that the fibres which contained 1% $CaCl_2$ yielded the optimum mechanical results with Young's modulus of ~1.9 GPa and ultimate stress of ~80 MPa compared to fibres produced with other components; however, they also demonstrated the highest degree of swelling at ~540%. In total, coaxial fibres of Chit/Alg exhibit a 260% increase in ultimate stress and more than 300% enhancement in the Young's modulus compared to their alginate counterpart. As expected, coaxial fibres had thermal properties and total weight loss occurring between the values observed for chitosan and alginate which contain about 60–70% wt. of the chitosan core and 30–40% wt. of alginate sheath when dried completely calculated from TGA results.

The biocompatibility of Chit/Alg fibres was demonstrated *via* cytotoxicity assays using two types of cells, human and murine myoblasts, in which cells were shown to be well attached to the fibre surface while having normal morphology, suggesting that the fibres showed no toxicity affecting the cell survival. Moreover, the use of these coaxial biofibres as delivery platforms have demonstrated significant improvements with regard to releasing the dye molecules permitting the release to be achieved in a more controlled manner compared to alginate fibres. These results, together with the cytocompatibility showed by them previously showed by them suggesting those as promising candidates for developing novel kinds of 3D bioscaffolds in drug release studies or tissue engineering.

Subsequently, GO was also inserted into the core of coaxial fibres to enhance the electroactivity of the conductive core and improve the mechanical properties of the fibres whilst maintaining good biocompatibility and cell adhesion of the scaffold by using a hydrogel polymer as the sheath. Use of ultra-large GO sheets has enabled development of a wet-spinning process for incorporation of a continuous core into fibres which is then easily convertible to electrically conducting graphene using facile thermal or chemical treatment methods. Chit/Go and Alg/GO coaxial fibres were successfully spun in a 1 M NaOH and 2% (w v^{-1}) coagulation baths, respectively. As observed from SEM images, while the GO in the core tends to blend with the alginate sheath in Alg/GO fibres to produce a final hollow-shaped structure, both chitosan and GO were held in their initial places after fibre formation in Chit/GO fibres.

Mechanical property results obtained from stress-strain curves for as-spun rGO and coaxial Chit/GO fibres showed a significant increase in robustness of the latter by about 33 times up to ~10.6 GPa. Coaxial fibres also revealed an ultimate tensile stress of ~257 MPa and elastic modulus of ~9.7 GPa, improvements of 8500 and 3350%, respectively, compared to the graphene oxide fibres alone. These increases were accompanied by an enhancement in elongation at break from ~1.5 to 3.1% (about 2-fold) in respect to the neat GO fibre. These superior mechanical property outcomes were shown to be in large part due to the major role played by the chitosan sheath as a result of the creation of a strong, homogenous interfacial connection between chitosan and graphene oxide. The average electrical conductivity of rGO fibres was measured to be ~48 S m^{-1} (an improvement of more than 2.5 times compared to ~19 S m^{-1} for GO fibres) after chemical treatment with ascorbic acid.

Both rGO and reduced Chit/GO fibre, demonstrated reasonable electroactivities. As calculated from CV curves, rGO fibres showed a specific capacitance of 38 ± 0.65 and 31.7 ± 1.5 F g^{-1} at a υ of 50 mV s^{-1} in PBS and aqueous 1 M NaCl solutions, respectively. Chit/GO fibres had specific capacitances of 22.2 ± 1.1 and 27.8 ± 1.5 F g^{-1} in PBS and aqueous 1 M NaCl solutions, respectively. The lower efficiency of Chit/GO fibres compared to that measured for rGO fibres might be as a result of a restricted access of the counterions to the conductive core due to the barrier created by coverage with the insulating hydrogel layer. The behaviour of two cell types, mouse fibroblast line (L-929) and rat pheochromocytoma line commonly used as a model of neural differentiation (PC-12), has also been investigated on GO and Chit/GO fibres. For both cell lines, cells were shown to adhere to the fibre surfaces with greater than 99% viability—results being very similar on the Chit/GO and rGO fibres. However, they will likely require some optimisation for any neural application since the neurite lengths obtained were shorter than what would normally be expected for the relatively long period allowed for differentiation.

PEDOT:PSS as one of the most highly conductive, processable, environmentally stable and biocompatible conducting polymers, appeared to be a very suitable candidate for development of coaxial electroactive fibres. Therefore, in order to improve the electrochemical properties of previously produced coaxial fibres, the GO core was substituted with PEDOT:PSS. The high dispersibility of PEDOT:PSS in water enabled the preparation of biocompatible, well-dispersed and homogenous spinning formulations from which Chit-PEDOT:PSS and Alg-PEDOT:PSS fibres were continuously fabricated successfully. Addition of 10% (w v^{-1}) PEG into the PEDOT:PSS core was found to have a negligible effect on the spinnability of this formulation, whilst affording the consistent and crack-free spinning of a core component that could be entrained consistently without break into the sheath. Interestingly, despite the absence of a reaction between alginate and GO in Alg-PEDOT:PSS fibres, PEDOT:PSS does not show any miscibility with alginate as observed when GO was exposed to alginate. This may arise from its higher viscosity (~6.7 Pa s) compared to GO (~1.5 Pa s) as a result of the presence of internal chain entanglements due to its polymeric nature. Exposure of oxidant-enriched coaxial fibres to an atmosphere containing the pyrrole monomer spontaneously initiated a polymerization reaction which resulted in the creation of a dark layer of PPy polymer on the surface of Chit-PEDOT:PSS (+PEG) fibres. The same polymerisation of pyrrole could not be initiated on the surface of Alg-PEDOT:PSS (+PEG) fibres.

Mechanical property results obtained for as-spun coaxial fibres showed a significant increase in robustness for them compared to the results previously reported for PEDOT:PSS fibres by others (~3.3 GPa). Young's moduli of fibres were calculated to be ~3.9, ~5.1 and ~4 GPa for Chit-PEDOT, Chit-PEDOT (+PEG) and Alg-PEDOT (+PEG) fibres, respectively. The results indicate that the presence of a hydrogel sheath plays a major role in increasing the Young modulus. Deposition of the third layer (PPy) was shown to decrease Young's modulus to ~2.2 GPa. Furthermore, Chit-PEDOT:PSS (+PEG) fibres were shown to be able to withstand higher stresses before breakage (~160 MPa) in the elastic regime compared to Alg-PEDOT:PSS (+PEG) fibres which showed ultimate stresses of ~94 MPa. On the other hand, Alg-PEDOT:PSS (+PEG) fibres revealed higher strain percentages before fracture with ~16% elongation before breakage. Triaxial fibres of PPy- Chit-PEDOT (+PEG) fibre were shown to have an ultimate tensile stress of ~64 MPa with ~3.5% ultimate strain.

Reasonable electroactivity was observed in CV results from coaxial fibres in both aqueous and organic media. The curves showed obviously larger CV areas, and correspondingly, higher specific capacitances in 0.1 M $TBABF_4$ than those of the CVs obtained in PBS solution. Specific capacitances of 19 ± 0.65, 11 ± 1.2 and 9.2 ± 0.8 F g^{-1} at a υ of 10 mV s^{-1} were calculated for PEDOT:PSS (+PEG), Chit-PEDOT:PSS (+PEG) and Alg-PEDOT:PSS (+PEG) fibres, respectively in PBS solution (based on the calculations in Sect. 2.2.1.11). Much higher capacitances of 29.7 ± 1.5, 21 ± 0.2 and 13.05 ± 0.6 F g^{-1} were measured for PEDOT:PSS (+PEG), Chit-PEDOT:PSS (+PEG) and Alg-PEDOT:PSS (+PEG) fibres, respectively in 0.1 M $TBABF_4$. With a view to optimising the performance of any potential battery structure, the ionic conductivities of the polymer electrolytes (hydrogel layer) were recorded and optimised by measuring the impedance behaviour using varying salt contents. Sodium chloride at 1% (w v^{-1}) was chosen as the desired salt and concentration to be added to both hydrogels and resulted in the highest ionic conductivity in chitosan (~15.1 mS cm^{-1}) and alginate (~22.7 mS cm^{-1}) while also displaying lower standard deviations compared to lithium chloride.

The biological tests revealed that on coaxial Chit-PEDOT (+PEG):PSS fibres the myoblast cells attached, differentiated and proliferated showing the typical cell morphology of myoblast cells. Nevertheless, it was also observed that cells did not show any tendencies toward attachment to coaxial fibres coated with alginate and attached and differentiated much better on coaxial Chit-PEDOT:PSS (+PEG) fibres than on the Alg-PEDOT:PSS (+PEG) ones. This combination of high electroactivity, remarkable Young's modulus and yield stress combined with the greater biocompatibility of these fibres imply that they hold great promise in applications such as drug delivery, batteries, power supply and tissue scaffolds in the near future.

6.2   Comparison of Fibre Properties

Electroactive biomaterials are a part of a new generation of "smart" biomaterials that allow the direct delivery of electrical, electrochemical and electromechanical stimulation to cells. Multifunctional hybrid fibres are one of the recently developed categories of smart biomaterials which have revealed a number of characteristics that show great promise for their potential use in a broad range of devices and applications including microelectrodes, neural implants, controlled release host, tissue-engineered scaffolds or actuators. The final physical properties of these fibres may be easily tailored to meet the essential requirements for a particular application. For all of those functions, there is a necessity for the fibres to possess high strength—not only to be able to survive the processing operations such as knitting or weaving but also vital to its continuing success in vivo, particularly for long-term applications in things such as implantable devices. Other key properties, such as microstructure, electroactivity, conductivity, degradation rate and elasticity of the fibres could also be regulated for any particular application. Charts of some of those key properties of the fibres such as tensile strength, modulus, elongation at break and electroactivity are presented below, allowing comparison of the different properties of the fibres prepared throughout this research.

6.2.1 Mechanical Properties

Mechanical properties of all fibre types produced in Chaps. 3–5 are depicted in Fig. 6.1. Enhancement of Young's modulus and tensile strength were observed in most of the coaxial fibres (except Chit/Alg) compared to their individual components. These enhancements were clearly higher for coaxial fibres containing reduced graphene oxide compared to those with PEDOT:PSS. While Young's modulus and tensile strength of Chit/Go are higher than the values obtained for Chit/PEDOT:PSS (+PEG) and Alg/PEDOT:PSS (+PEG), it was observed that the elongation at breakage is significantly lower. To take advantage of the strength and stretchability characteristics of both of these potential conductor elements, it might be useful to try a blend of them as the core component.

6.2.2 Electrochemical Properties

Figure 6.2 compares the electrochemical properties of all fibre types produced in this study. Generally speaking, solid conducting fibres demonstrated higher electroactivities and achieved higher specific capacitance compared to that of coaxial fibres. The poorer performance of coaxial structures could be attributed to the more restricted access of the counterions to the conductive core since it is covered with an insulating hydrogel layer.

Moreover, according to the results in Fig. 6.2, incorporation of reduced graphene as the core component gave rise to the specific capacitance values in all fibres. Another interesting investigation that suggests itself from looking at this graph would be the effect of the presence of a strong chemical interconnection between two layers in the coaxial structure in achieving high capacitance values. It was obviously seen in the results from Alg-PEDOT (+PEG) fibres. Significantly lower specific capacitance was

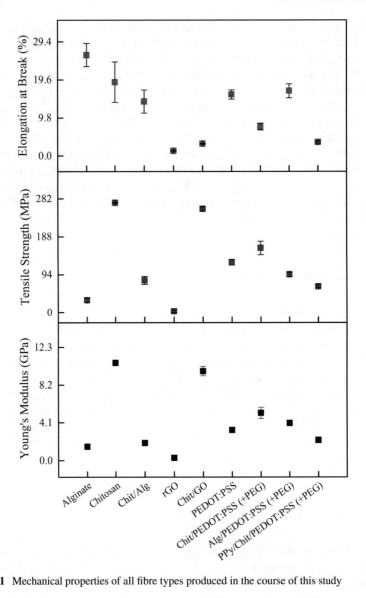

Fig. 6.1 Mechanical properties of all fibre types produced in the course of this study

calculated for Alg-PEDOT (+PEG) fibres compared to that of Chit-PEDOT (+PEG) fibres despite their higher ionic conductivities in wet-state. This could be attributed to the lack of any bonding between the alginate and the PEDOT:PSS.

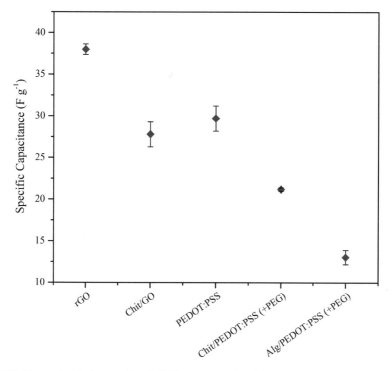

Fig. 6.2 Electrochemical properties of all fibre types produced in the course of this study

6.3 Recommendations for Future Work

A range of multifunctional multiaxial fibres has been developed for a wide range of purposes. The main aim of this thesis was to establish a wet-spinning process to develop three-dimensional coaxial and triaxial electroactive fibres which contain a conductive core enclosed in an appropriate biomaterial sheath which enhances cellular adhesion and proliferation properties. This goal was successfully achieved using several organic conductors and hydrogel biomaterials. However, like any other research, there are many promising areas still to be pursued in future studies which could not be overcome in this one.

- Battery discharge tests are required to be carried out to test and regulate the performance of the triaxial fibres as potential batteries. Time restrictions did not allow this to be completed as part of the current thesis.
- Fabrication of triaxial fibres using a one-step wet-spinning method *via* a triaxial spinneret is also suggested. Using this approach will probably improve the attachment of the various battery layers as well as the thickness of the outer layer, resulted in better battery performance altogether.

- It would also be beneficial to try other approaches for the production of triaxial fibres such as either a combined wet/electrospinning technique or utilising a twisting method to place the second conductor on the surface of coaxial fibres electrochemical deposition method were not possible to investigate due to the time limitations. Comparison of the properties that each of these methodologies produced would be very useful in the successful production of an optimal battery device.
- Also, it would be desirable to fabricate these functional fibres into devices with various spatial patterns using fabricating techniques such as weaving, knitting and braiding methods. These textile structures may then find use as templates for the regeneration of fibre-shaped functional tissues that mimic muscle fibres, blood vessels or nerve networks in vivo.

Printed in the United States
By Bookmasters